BRAND
PLAN Rx

Markus Saba & Hilary Gentile

BRAND PLAN Rx

The Marketer's
Guide to Building a
Thriving Health and
Wellness Brand

PAGE TWO
BOOKS

Cataloguing in publication information is available from
Library and Archives Canada.
ISBN 978-1-77458-103-2 (paperback)
ISBN 978-1-77458-104-9 (ebook)

Page Two
pagetwo.com

Edited by Scott Steedman
Copyedited by Alison Jacques
Proofread by Alison Strobel
Cover design by Fiona Lee
Interior design by Setareh Ashrafologhalai
Interior illustrations by Michelle Clement

brandplanrx.com

CONTENTS

ACKNOWLEDGMENTS

WE SOUGHT COUNSEL from veterans in our industry and perspectives from fresh eyes.

First, we would like to thank Rachel Reckner, a colleague and dear friend who helped us with the first draft of this book, bringing her skills, creativity, writing ability, and, most importantly, knowledge of the industry to our work.

Second, we extend thanks to Tim Calkins—author, consultant, and clinical professor of marketing at Northwestern University's Kellogg School of Management—for his pioneering work on branding. In particular, his book *Breakthrough Marketing Plans* provided us with fertile soil in which our ideas took root.

Third, we warmly acknowledge the team at McCann Health, with whom Hilary Gentile collaboratively developed several concepts in this book. Hilary credits McCann's dynamic work environment which encourages the cultivation of innovative approaches to healthcare communications.

Finally, we would like to thank the thirty students from the University of North Carolina Kenan-Flagler Business School's "Healthcare Brand Plan" class of 2020, who used this book as a pilot in the course. They provided us with vital insight and an understanding of what worked and what didn't; how best to explain concepts; and what constructs, models, and examples were helpful, and which ones to leave out.

PREFACE

I didn't have time to write a short
letter, so I wrote a long one instead.

MARK TWAIN

MARKETERS NEED TO make choices, to distill the essence of their brand. This process of simplifying is crucial to building a brand plan. Of course, a diligent marketer gathers all the relevant information, data, and evidence about a category and brand before they create a brand plan. But to be successful, they must then digest all these inputs and thoughtfully choose what matters most to drive their brand's success. And the simple map they produce must be cohesive, with well-defined links between all the parts, so it is crystal clear how the brand will achieve success.

Brand Plan Rx will help you do just that. This book provides you and your teams with an aligned process and methodology tailored to the health and wellness industry. Proven constructs give you the tools you need to make those critical choices and ensure that your plan is cohesive. Plenty of real-world examples and practical advice bring those concepts to life.

We have decades of experience developing, writing, presenting, reviewing, approving, declining, and implementing brand plans in the healthcare field over the course of our careers. Literally thousands of plans, in numerous therapeutic areas, across the globe and with a variety of companies. Together we have deep marketing expertise with a global footprint, having lived and worked around the world. We have experienced brand plans from the client side as well as the agency side. We have developed brand plans for small assets, rare diseases, biologics, chronic disease, various therapeutic areas, and the biggest and most successful brands in the industry. We have done this with pharma, providers, payers, and numerous other health and wellness companies. We have seen some brand plans sail through the approval process and seen others struggle to achieve momentum. *Brand Plan Rx* is a compilation of all we have learned after decades of successes and failures.

A few years back, Markus developed a course titled "Healthcare Brand Plan" at the University of North Carolina Kenan-Flagler Business School for the MBA class within the healthcare concentration. He collaborated with Hilary on the content, best practices, examples, principles, and concepts. In fact, she became a regular guest lecturer in the course, which was an instant hit. The students loved the practical learning, which they immediately applied in their jobs.

For this course, we found a couple of excellent textbooks that explained the marketing concepts and the brand plan process. However, none of them were customized for healthcare or took into account many of the industry's unique conditions and nuances. We also used books that talked about the pharmaceutical industry, but those only focused on pharmaceuticals, particularly on advertising, not the brand plan. Other health and wellness marketing books focused too much on the science, virtually ignoring the marketing of that science.

As we continued working with clients, taught courses, and conducted workshops, people kept asking us what resources would help them write better brand plans for the health and wellness industry, especially in pharma. We struggled to give them a good answer or to name a definitive title, a manual that would inspire their thinking and outline best practices. We realized that there was no simple go-to manual for our industry, no one source that teaches how to write a brand plan by detailing the best processes and practices and sharing real-world success stories and failures specific to health and wellness.

We knew we needed to create a simple cohesive road map to brand planning in the health and wellness industry. Our course bridged the gap between these resources. Over the years we perfected our methodology and pedagogy. Every year, we had students tell us, "Hey, you should write a book explaining how to write a brand plan—then we wouldn't have to buy three books and do all the interpreting and adjusting." We heard the same request from industry. Our clients were asking us for additional resources and direction. They were asking for the art and method of brand planning to be consolidated into one reliable source specific to our industry. We eventually felt we had the process mastered and began writing.

Brand Plan Rx is the marketer's guide to building a thriving health and wellness brand.

This book was developed specifically with the health and wellness industry in mind. We have taken traditional marketing principles and modified them to be more appropriate to the life sciences, incorporating the science, innovation, regulations, providers, payers, prescribers, and patients—all the elements that make marketing in the life sciences unique and requiring a slightly different approach at times. As much of our experience comes from the pharmaceutical world, you will see many pharma examples. However, we have applied our principles and methodology to other sectors of the health and wellness industry with great success in our careers and show how you can too. In addition to pharmaceutical products, our techniques will help you write brand plans for any product or service that will benefit the health and wellness of a consumer.

After we developed our first draft of this book, we thought it would be a good idea to test-drive it. Curious marketers that we are, we thought of a creative idea—the ultimate in market research. We gave the manuscript to a class to use as a textbook. In the fall of 2020, the UNC Kenan-Flagler Business School "Healthcare Brand Plan" class got to use the manuscript of *Brand Plan Rx* for the first time. Students gave us feedback, suggestions, and ideas, and pointed out areas for improvement. The book you are reading is the result of that live test and real-world experience.

A Unique Industry

As we have alluded to, health and wellness brand plans are different from those in other industries, most notably the consumer packaged goods (CPG) and retail industries. While there are a number of similarities, the differences are significant. First and foremost, health and wellness is highly regulated—much of what you can say and do in CPG and retail is not allowed there. You see this with prescription medicine ads on TV, which, due to "fair balance," have to list all possible side effects. You don't see this with chips or candy bars (though you probably should consider some of the health concerns of these products). The approved label found in your drug's package insert states what you are able to claim and not claim promotionally.

In most consumer marketing, you also market directly to the end user, the person who will buy and use the car, shoe, shampoo, or candy bar. Pharma is more complex, with many more stakeholders to consider, as it is a B2B2C (industry to prescriber to patient) business. What and how you communicate to patients is different from what and how you communicate to doctors, which differs again from what and how you communicate to payers. To complicate matters more, the group and/or person making the brand decision is usually not the group and/or person paying for the brand, nor the group and/or person using your brand—often these are three separate entities making decisions that are not always aligned in the same manner. You need to consider all of these stakeholders simultaneously while developing your brand plan.

Research and development (R&D) in health and wellness is also different from any other industry. Pharma brands have patents that expire about ten to twelve years from launch, cost billions of dollars, and have success rates of 1 in 10,000. This means that your adoption curves and share of market (SOM) growth need to look very different from those in CPG and retail.

Finally, the motivations of the brands are very different. People *want* to buy CPGs and retail, because these products celebrate their "self." They willingly buy cars, clothes, and food as a way of expressing their aspirations. However, no one *seeks* to buy a healthcare product— they do so because they *need* to. Patients often hate that they need to take medicine, and many try to hide the fact that they have a disease and are on medication, in a protection of their "self." In pharma, marketers need to investigate the context, the self, and interdependent trust factors along the journey of the disease to uncover the needs that drive brand choice.

How This Book Is Organized

This book is designed so that you can easily move from one chapter to the next and then circle back when needed. It is organized to take you methodically through the road map step by step, with one chapter for each step. As you go along, keep referring back to the Brand Plan Rx Choice Map. The first four steps in it are shown as an iterative process, and as you learn about your brand you may need to update your customer insights, enhance outputs of the brand assessment, and so on. The second half, the last four steps, is designed in a more sequential manner, meaning that one set of decisions builds on the previous set.

In each chapter we introduce the concept, discuss how to apply it, and then look at case studies. We start by helping you gain an in-depth understanding of the topic by reviewing definitions, theories, and context. You will be shown constructs and templates that we have found to be effective over the years. This will enable you to bring your ideas to life, and show you how to apply concepts to the brand plan you are working on today, questions to consider, and challenges to address.

The case studies come from our decades-long experience working in health and wellness all over the globe. We are firm believers in the surgeon's creed: "see one, do one, teach one." You only truly begin to understand a process and appreciate the nuances once you have had to teach one, be it by leading a team or by lecturing in a classroom. We have compiled all those experiences into a concise, roll-up-your-sleeves go-to guide. The best part of living through all of these opportunities is being able to pass on what is really needed to build a plan that works.

Practical Tips

In addition, where applicable we provide "Watch Out" and "Let's Get Real" sections to help you avoid making mistakes that we and others have made. Watch Out sections are "minefields" we want to warn you of in advance. In the Let's Get Real sections we apply our real-life experience to challenge conventional wisdom. We've learned that the conventional path is not always the most effective route.

Building a brand plan is a complex and messy process. You will come up with many ideas and slides that don't end up in your plan. They may be left out of the story; however, they were important building blocks and thought processes that allowed you to get to the brand plan. And that's okay.

Modern Marketing

All through this book you will see mentions of "modern marketing": new approaches, philosophies, methodologies, and insights that modern marketers are applying. We have written this book with the core fundamentals to build a choice map and a cohesive map to deliver the right Rx, or prescription, for your brand. There are promising new ways of approaching many of the inputs to these time-tested foundations. With COVID-19 accelerating our openness to inventive thinking and data-led decision making, we would be remiss not to highlight areas that are nascent yet very effective in unlocking the maximum potential of your brand.

Moving Forward

The Brand Plan Rx Choice Map is your ultimate guide to making decisions for your brand's success. Only when you are confident you have a cohesive, well-thought-through, data-led story for your brand will you able to fill in the final draft of your Brand Plan Rx Cohesion Map. All the input is necessary to tell a succinct story of how you will drive success for your brand. Which brings us back to Mark Twain, who expressed that it takes a lot more cognitive determination to write a short letter than a long one. Make sure your snapshot is laden with thoughtful choice built by diligent interrogation of your brand's opportunity in the market.

Key Questions

At the end of each chapter you will see a list of key questions. These are questions that we would ask ourselves if we were developing the brand plan. Or they could be considered questions we would ask of you if you were presenting your plan for approval.

INTRODUCTION
THE BRAND
PLAN Rx PROCESS

Key Learning Points

Brand Plans: Process and Content

Brand Plan Rx Choice Map

Brand Plan Rx Cohesion Map

Launch Brands vs. On-the-Market Brands

Outside-In Thinking and Inside-Out Thinking

Brand Plans: Process and Content

Writing a brand plan is the single most important exercise for a marketer. Without one, you don't have a brand. Companies are successful because of their brands, and behind every successful brand is a brilliant marketer. Marketing is all about making choices and getting those choices on paper in a cohesive manner that spells out how the brand will be successful. It is easy to pontificate about what it will take, but writing a succinct story of how a brand will fulfill its potential is a critical skill for every marketer.

The brand plan is a road map for all internal stakeholders. It clearly outlines what it will take to drive the business forward. It is the strategic plan that sets the course, spells out the goals, aligns the organization, sets priorities, and states precisely the actions to be taken. The brand plan determines where you are going and how you will get there.

Creation of the brand plan is typically an annual exercise. It takes up an enormous amount of time, energy, and resources. Some pharmaceutical firms start brand planning for the next cycle as soon as the current plan is approved. During the review and approval process, the strategy is pressure tested, tough choices are made and defended, novel ideas are introduced, credit is given, and accountability is assigned. The brand plan can make or break a career.

All our experience and learning have led us to two cardinal tenets:

The first is **choice**. If you can't summarize a brand plan on one page, you haven't made the tough decisions to drive success. The only way to obtain that simplicity is by making clear choices.

The second is **cohesion**. It all has to fit together. There should be a "golden thread" that aligns the plan to a common understanding. Your analysis should lead to clear choices and your corresponding strategy and tactics should all be connected.

To bring choice and cohesion to life in the brand planning process, we have two unique and proven constructs. Together, they are designed to anchor your methodology and to make sure you have made choices and your brand plan is cohesive.

Brand Plan Rx Choice Map

The Brand Plan Rx Choice Map has eight critical thinking inputs: environmental pulse, customer deep dive, brand assessment, positioning, objectives, game plan, solutions, and writing the plan. The inputs have unique approaches and customized tools to guide you in your decision-making journey.

The Brand Plan Rx Choice Map is broken down into two phases. The first phase is an iterative process that includes steps one to four: environmental pulse, customer deep dive, brand assessment, and positioning. This is the input and analysis phase. During it you are conducting research and analysis that unlocks the *who* and *why*. Once this is complete, phase two begins and you are ready to follow steps five to eight. The second phase is a more linear process in which one decision is linked to the next. You'll make decisions in a stepwise fashion, first determining objectives, then defining your game plan and developing solutions, and finally writing the plan. During this phase

BRAND PLAN Rx CHOICE MAP

| WHO & WHY | WHAT & HOW |

Environmental Pulse → Customer Deep Dive → Positioning + Objectives → Game Plan / Solutions / Writing the Plan

Brand Assessment

you are making decisions that answer the *what* and *how*. All the input, analysis, and information of the first phase leads to decision making in the second phase. After completing both phases, you have finished the steps on the Brand Plan Rx Choice Map.

Brand Plan Rx Cohesion Map

The final step of your Brand Plan Rx Choice Map is writing the plan. Here, all your research and critical thinking comes together in one cohesive place to summarize your plan on one page. This plan-on-a-page is your Brand Plan Rx Cohesion Map. It contains every component required to arrive at a cohesive and clear brand story. This is just the prescription your brand needs.

The Brand Plan Rx Cohesion Map is purposeful. Being able to connect inputs to outputs and clearly synthesize how your brand will succeed on one page gives you a north star, a clear road map to use as your sounding board for all decisions you make for your brand. We have purposefully developed a "cohesion map" so you can take a step back and determine that the choices you have made for your brand are interlinked. This is one of the most important learnings from our years of brand building. The Brand Plan Rx Cohesion Map is a tool that forces you to determine whether your choices are built with the overall picture in mind. As you master synthesis and clarity, you will ensure your brand is set to make a marked difference in people's lives. The process of building the cohesion map enables a snapshot summary and synthesizes your complete brand strategy into a simple message you can share with your team, cross-functional members, and busy executives, in an elevator or on a poster.

The Brand Plan Rx Cohesion Map makes your brand plan coherent, digestible, and memorable. It shows the method to your thinking and the choices you have made to create a distinctive and significant opportunity for your brand. Limiting the real estate to one page reinforces the need to make trade-offs and drive choices. The flow of the map should intuitively lead the reader to see the "golden thread" of the plan. We are always asking ourselves, would someone understand our intent if they didn't know the category? Do all the inputs add up? Are they connected and building to our recommended strategies and initiatives?

Ultimately, brand plans are about making choices by assessing the environment, turning data into recommendations, and aligning the organization and resources. The methodology in this book will help you accomplish exactly that—all on one page.

The cohesion map summarizes your key decisions. It allows you to see in a snapshot if your plan is aligned and if each decision leads logically to the next. For example, do your key takeaways from the environmental pulse and your key insight align with your brand objectives? Do your brand objectives connect logically to your game plan?

An effective way to use this tool is to work it backwards. Start with a solution and make sure that it supports and aligns to a specific strategy in your game plan. From there, make sure that your game plan will result in achieving your brand or customer objective and that your objectives are aligned with what you have learned about your

BRAND PLAN Rx COHESION MAP

WHO & WHY	WHAT & HOW

MARKET, CUSTOMER & BRAND UNDERSTANDING	BRAND & CUSTOMER OBJECTIVES	GAME PLAN	SOLUTIONS

Environmental Pulse

Customer Deep Dive

Brand Assessment

Cohesion Statement

Strategy #1 → Big Idea

Strategy #2 → Big Idea

Strategy #3 → Big Idea

← Brand Positioning Statement →

environment, your customer, and your brand. Finally, ensure that your brand strategy supports your positioning—your "true north." The Brand Plan Rx Cohesion Map is a synthesis of your thinking, analysis, and decision making, in a simple one-page snapshot that you can easily refer to, communicate, and remember.

As you go through this book and develop your brand chapter by chapter, keep the Brand Plan Rx Cohesion Map in mind. Consider how you would summarize and synthesize the work you have completed. Consider if there is alignment from one understanding of your customer and brand to the next. Consider if there is alignment from one decision to the next. In chapter 8, near the end of this book, we will focus on the actual development of the Brand Plan Rx Cohesion Map and explain in detail how to write it. For the time being, simply keep it in mind and know that each chapter is designed to help you create your Brand Plan Rx Cohesion Map.

Launch Brands vs. On-the-Market Brands

You might spend more time on certain parts of the map than others, depending on whether you are working on a launch brand or one that has been on the market for a while. For a new brand you are completing analysis, obtaining information, and making decisions for the first time. You have to build the foundation of the brand, such as its positioning (which will be enduring) and the branding elements. Brand foundations are the crucial elements you build before launch. Some should last for the life of your brand. These brand foundations do not change from year to year; rather, your understanding gets enriched and the implementation is fine-tuned.

For an existing brand, your positioning has already been developed. Only in very specific circumstances will you need to revisit a brand foundation step and make any major overhauls or changes. These include a new indication or line extension (NILEX), a new delivery device, new regulations or guidelines that change the landscape of the market, or perhaps a new competitor with a truly innovative or disruptive brand. Under these circumstances you should review your brand plan and strategy and determine if significant changes should be made to them. The resulting conclusion is rarely a change in your brand foundation (more often it's a change in messaging), but evaluating the situation is the responsible thing to do.

For an existing brand you are pressure testing your assumptions against changes in the marketplace or evolving customer expectations to drive further success. The brand plan steps in the process map should be reviewed on an annual basis with the expectation that they will need to be enhanced significantly. Plans should be written with a two-year time horizon. The first year of the plan should be clear and understood in detail, since the team will need to deliver on it immediately. Year two is a simple extrapolation of year one. It aims to plan for the second year of success, as some strategies and initiatives take more than twelve months to have an impact. Your ideas, innovations, solutions, and programs should not be limited to a twelve-month cycle simply because of the calendar. It all depends on what time of the year initiatives are implemented and how long it takes for them to show impact.

The time horizon of a brand plan authored by a global team rather than a country team (sometimes called an affiliate or a market) may be longer. The two-year time horizon referenced above is a good rule of thumb, but many global teams choose to take an even longer view, especially when looking at life-cycle planning. They are usually accountable for the brand positioning and therefore must take into account NILEX on the horizon, updated core claims, scientific innovations, and so on. They may also need to reference scenario planning for prospective environmental evolution. Regardless, a global team needs to understand core market insights when developing the global plan (and if possible, leverage country input to co-create it). The affiliate or market teams are more focused on implementation, so a one-to-two-year time frame is appropriate for planning purposes.

Outside-In Thinking and Inside-Out Thinking

Another key concept you will find throughout the book is that of outside-in thinking and inside-out thinking.

We mentioned that there are eight steps to our road map. The first two, environmental pulse and customer deep dive, require outside-in thinking. This is an important perspective to understand. The best marketers embrace it in everything they do and continuously keep it top of mind along the entire brand plan process. Step three, brand assessment, is when inside-out thinking begins to be applied.

Understanding your environment begins with a deep understanding of your market and your customer. The dynamic of how those two elements interact is crucial to understanding your key insights. This all begins with an outside-in mindset. This approach starts with an external market orientation and dives deep into market trends and customer insights before developing the brand strategy. Start with the external market orientation and study the trends affecting the healthcare system, disease, category, providers, payers, caregivers, and ultimately the patient in order to develop your brand plan strategy.

George Day and Christine Moorman bring this concept to light in their book, *Strategy from the Outside In*. They explain it by saying, "Outside in means standing in the customer's shoes and viewing everything the company does through the customer's eyes."[1] They illustrate this idea with examples across various industries—but, interestingly, few from pharma. As healthcare and wellness are complex and highly technical industries, it is easy to understand that, for example, pharmaceutical companies begin with the science and the data rather than with the customer. It is, after all, the scientific data that allows for innovative solutions.

Many senior executives are scientists, researchers, or doctors, and companies invest billions of dollars in research and development to fulfill unmet needs of disease or betterment in our health and wellness. Therefore, a keen focus on the science makes sense. The pharmaceutical industry has made great progress in embracing this outside-in concept to influence brand strategy in recent years, but is still behind other sectors. We, as an industry, have a responsibility to gather a deep understanding of the environment and insight from the customer's perspective. Employing the C-suite and other functions such as research and development, medical affairs, and regulatory affairs to understand and prioritize this outside-in thinking is paramount. Brands that understand the market environment and customer best will have a clear competitive advantage in the marketplace.

1

ENVIRONMENTAL PULSE

Key Learning Points

Relevant Trends

Headwinds and Tailwinds

Defining the Era

Competitive Analysis

BRAND PLAN Rx CHOICE MAP

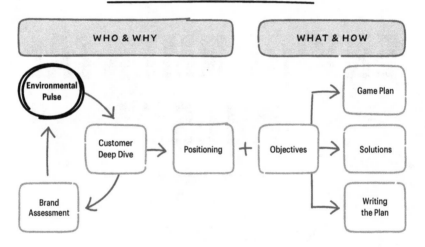

TARTING WITH the environment, the context, the culture, the world your brand will be living in is of critical importance. Understanding the market means looking at current solutions, competitors, trends, and changes. You need candid and honest information and feedback. You need pushback from your customers. You need objective analysis about the market in which your brand will compete.

This environmental pulse exercise will give you a very robust assessment, one that you will need to communicate with alignment and political deftness regardless of the outcome. Whatever the scenario, your approach to building a brand plan stays the same. Step one begins with an external analysis regardless of whether your product is a new brand, in-licensed, or an already on-the-market asset.

Keenly analyzing your market reveals the forces and dynamics that will enable the health of your brand. If you have an accurate and deep understanding of your environment—your fingers on the pulse—then your brand plan will be designed with a perceptive context in mind. Alternatively, if you do not fully understand the environment, you risk missing the mark.

The pharmaceutical industry is full of brand successes and failures. The brands that have been successful seemed to "just get it." They had a deep understanding of the environment and the marketplace, understood the trends, and anticipated the future. When researching why brands failed, in contrast, one common theme is that they simply didn't understand their environment. They missed the mark based on this fatal flaw, one that in retrospect seemed very obvious: If you are launching a new brand, the environmental analysis is extremely important. There is much work to be done, including market research, data analysis, and secondary research. This often requires building your understanding from the ground up. The benefit to this is that with fresh eyes and an open mind, you are likely to see things without bias and be able to anticipate the future.

If you are completing your brand plan for an existing brand, this exercise could be more challenging. You need to validate your existing understanding of the environment. You and your team are coming into this with established understandings, experiences, and biases. Driving a customer to shift a belief, to move from an existing thought or understanding, is much more difficult than building on an existing belief system or unlocking a novel need that answers a gap in the market. Marketers can become entrenched in what they understand about the market. Their experience becomes the reality for the entire market.

The first step in understanding your external environment is the process of identifying all the external elements that can affect your brand's performance. This entails assessing a number of aspects in an organized and structured manner. Begin with understanding relevant trends in your category and/or disease state, by applying outside-in thinking.

Outside-in thinking is critical when building your plan. The understanding of "where you are now" and "where you will be" is the foundation of your brand plan. It is helpful to begin by understanding which era you are in.

Watch Out

One of the first potential disconnects can occur when senior leaders begin to hear your in-depth market assessment and customer feedback after a period of relative silence that has followed the initial commercial decision or acquisition of the asset. The business development team or product development team has likely done a cursory assessment of the asset and will have already painted a picture of peak sales and the assumptions that underpin them. We have seen that marketing experts, who will be responsible for the commercial success of the asset, are not deeply involved at this early stage. Executives will have formed an opinion about the market, brand, and opportunity based on a quick, early, and high-level assessment. At times, the due diligence required to truly understand the environment, customer, and competitive frame has not been comprehensive. The team will have obtained some input and feedback from key opinion leaders (KOLs). While KOLs are important stakeholders with relevant and vital perspectives, they often do not reflect the "real world." They aren't usually in the trenches treating patients on a daily basis, handling access issues, managing problems the average clinician encounters on a daily basis.

This focus on internal thinking can also extend to your marketing team. We, as an industry, can spend years and millions of dollars working on a molecule to turn it into a brand and a commercial success. Understandably, we become attached. We start "drinking the Kool-Aid." In fact, some noncommercial colleagues wonder why we need all the marketing and brand planning inputs and iterations. This drug they've developed is so great, the benefits so obvious, the science so clear and innovative, that clinicians will be eager to prescribe it given the right medical education tactics. The clinical evidence clearly shows how the product improves patient outcomes. Doctors will know how and when to use it, they will be writing prescriptions for all the "right" patients, and the payers will happily ensure access and reimbursement at a good price and a healthy margin. So why waste all that money on marketing?

Not so fast. This internal, product-centric approach is the main reason that brands fail. Companies don't understand the marketplace: They believe their products are great and don't properly build the brand. They fail to adequately develop the market, educate prescribers, inform patients, or convince payers. The way to avoid this trap is to understand the business from the outside in—from the customer's perspective, not the brand's or company's perspective.

Relevant Trends

The relevant trends are the ones that enhance your understanding of your customers and the changes in your competitive landscape. Your brand plan should clearly identify a few key trends that are transforming the healthcare space and are relevant to the category in which your brand is competing.

It's important to look at both current and future trends. The last thing you want is to be blindsided by something on the horizon. How are your customers changing? How are they solving their problems as their needs evolve? What competitors are on the move? Are there innovators entering the market that will transform it? When you are building out potential scenarios, you need to also identify the factors that could change, just as we saw the HIV/AIDS market change over time. Monitoring these changes in real time will be important, so that the context of your strategies is clear. Keeping your fingers on the pulse of the environment is an ongoing and never-ending exercise.

Finding an efficient way to capture and categorize key information and learnings is important. You need to establish a process and a repository to capture various analyses. People move on and take their knowledge with them. New information comes out that confirms or contradicts previous understanding.

The number of trends you can consider is endless. The key is focusing on trends that are most relevant to your customers, their disease, and your brand. Here are some ideas to consider and stimulate your thinking:

- **Health factors:** demographics, culture, ageing, obesity, language, exercise, smoking, drug use and addiction, social determinants of health (SDOH)—income, food, housing, transportation, etc.

- **Healthcare system:** changes in policy, universal healthcare coverage, underserved populations, Medicare, Medicaid, regulations, Food and Drug Administration (FDA) policy, uninsured, overinsured

- **Health industry:** vertical integration, consolidations, insurance reform, scientific innovation, precision medicine, genetics, research and development, services

Once you've looked at the main trends, you should also look beyond the obvious. Go one step further and consider consumerism, sustainability, health of the healthcare workforce, and SDOH. The increase in consumerism includes such aspects as health literacy and care anywhere. Aided by advances in technology and digital health solutions, patients are acting more and more like consumers. They often have access to their health data at the same time as their health-care providers. Artificial intelligence (AI), wearables, telehealth, and digital apps have completely changed the patient experience and access to their information.

Concern about the sustainability of the US healthcare systems continues to be an emotional and hot topic. The debate around universal coverage and a single-payer system continues. A number of factors contribute to this highly complex issue. Increasing costs, fee for service, value-based care, more elderly and indigent patients, pricing transparency, increases in noncommunicable diseases (NCDs), and end-of-life care are just some of the things to consider.

The health of the healthcare workforce is another trend you should consider. About half of the doctors in the United States are over the age of fifty-five. Doctors are already experiencing some of the highest levels of stress and divorce, among other challenges. The debt and time required for medical school are contributing factors, along with the actual work environment. Ironically, the physical and mental health of many of our doctors is under strain, leading to tremendous burnout. A 2020 Medscape study of over 29,000 doctors in the United States revealed that 42 percent of them reported feeling "burned out" and more than 20 percent have had thoughts of suicide.[2] It's important to note that this study was fielded *before* the COVID-19 pandemic. One can only imagine how challenging the current healthcare working environment is. Given the high medical school debt load and the subsequent work stress, it should come as no surprise that we are facing a huge shortage of doctors and nurses with no relief in sight. This will impact your brand strategy in a number of ways, as the quality of care delivered, access to care, wait times, capacity, and availability continue to face real challenges. How and when you communicate to healthcare professionals (HCPs) will evolve and change.

Another set of factors to look at are the SDOH of your patient population: such things as economic status, social status, housing, food, education, transportation, and access to healthcare. Helping to treat those in need is an important issue. Many of these factors will affect your brand in different ways and to differing degrees. Analyzing them will give you insight into potentially relevant trends.

Modern Marketing

The COVID-19 pandemic is one of the major healthcare macroforces of our generation. It has affected all aspects of life and all industries. Marketers have had to add an additional layer of questions to adapt to the pivotal dynamics that COVID-19 has introduced. The impact on healthcare has been profound. Some of the changes will be temporary, some will be transformative, and some will result in adjustments that will eventually settle into a new normal. Others will be permanent. Healthcare providers on the front line will see permanent changes to previous best practices and conventional wisdom. The rapid uptake in telehealth is just one example. In the pharmaceutical industry, a number of brand plan assumptions that held true in 2020 might be challenged now owing to the impact of COVID-19. The degree of change may differ by therapeutic area and brand.

Things to consider include the following:

- changes in public policies, regulations, and government agencies;

- access to hospitals, clinics, offices, doctors, and staff;

- best ways to communicate with patients and HCPs;

- impacts on the business model;

- modifications to the payer space in both the private and public sectors;

- ability to conduct clinical trials and bring a new molecular entity (NME) and new indications and line extensions (NILEX) to market;

- priority of the disease your brand impacts relative to the new healthcare priorities; and

- safety and delivery of care, especially for patients with underlying health conditions and for whom telehealth is not an option.

These are just a few examples of trends that are impacting the environment your brand is occupying. If you did a brainstorming exercise, you could probably list dozens of such trends and factors to consider, from social determinants of health to advances in science and technology to regulations and guidelines. You must decide which ones are worth your attention and choose the ones that are most relevant for your brand.

Headwinds and Tailwinds

Once you have identified relevant trends, you then need to decide if they present a challenge for your brand or work in its favor. We refer to these trends as "headwinds" and "tailwinds." Headwinds are potential challenges, issues, and barriers. Tailwinds are movements or trends that keep things going forward in the right direction. Riding the momentum of a trend is much easier and less costly than fighting against it. Creating or changing a trend takes time, energy, and money and distracts from what you are trying to build with your brand. Joining a movement is also much easier than creating one from scratch. Finding a tailwind is a win for your brand.

CASE STUDY: NARCAN NASAL SPRAY

A great example of riding a tailwind is the launch of Narcan Nasal Spray, a rescue medication to save people from opioid overdose. It was introduced just as the United States was recognizing the crisis of prescription pill addiction, with overdoses becoming the leading cause of unintentional deaths in the country for people under the age of fifty in 2018.[3] The federal government launched an expansive program to fight the epidemic of opioid and heroin abuse. Millions of dollars were spent to address this national health priority, including federal funding for states to purchase and distribute the opioid overdose reversal drug naloxone (more commonly known as Narcan). Guidelines and funding required Narcan Nasal Spray to be available in numerous places, from hospitals to police cars, and enabled it to be dispensed at pharmacies without a prescription. This is the kind of tailwind that will set your brand on a clear path for success. Sadly, it is a tailwind that we wish did not have to exist.

CASE STUDY: PRALUENT

Let's contrast this with a brand that launched in the face of a con-
siderable, underestimated headwind: pricing and access pressure in
the cardiovascular space. Praluent is a PCSK9, a novel medicine at
the time of its launch in 2015 for the treatment of high LDL cholesterol. It seemed to have a tailwind in its favor: the tremendous
unmet need in this disease state. One in four Americans will die of
cardiovascular disease.[4]

The data for this drug was compelling, but the price tag sig-
naled that the companies that had brought it to market, Sanofi
and Regeneron, were seriously out of touch with the payer reality.
The list price in 2015 was $14,600 a year, at a time when generic
statins (lipid-lowering medications) cost about $40 a year. Statins
are proven to lower cholesterol, and the clinical benefit that Praluent offered versus that of the statins did not appear to justify
the enormous price difference. Given the cost, insurers could not
allow widespread use of the new medication. Some estimated
that if open access was granted to it, healthcare spending in the
United States would go up by $120 billion a year.[5] Not surprisingly,
insurance companies placed so many restrictions on the use of Praluent that it never came close to realizing its potential. Clearly, at
the very high list price, the amount of discount or rebate was not
enough to make the drug competitively priced at the formulary or
patient co-pay level. As of 2020, Sanofi has gotten out of the US
market for Praluent.

Defining the Era

Understanding how the market exists today, and how it has changed or will change over time, will help you identify opportunities. Discussing future scenarios is an important part of this process. Take the time to debate what would drive Scenario A or B in the future. What will stay the same, and what could change?

CASE STUDY: HIV/AIDS

To see how the understanding of a disease and the market shifts over time, let's look at an example: the fight against HIV/AIDS. This began in the 1980s, when the disease first appeared and diagnosis was tantamount to a death sentence. Thankfully, the introduction of antiretroviral treatments changed this, but they were very complicated and involved complex dosing regimens. According to the CDC's National Center for Health Statistics, deaths from AIDS in the United States peaked in 1995 at 41,699 and just two years later had fallen to 16,685 with the introduction of GSK's (GlaxoSmith-Kline's) AZT (Azidothymidine) medication Retrovir.[6] Shortly after that, the market matured to focus on treatment convenience while continuing the search for a cure.

As science and research continued to evolve, so did the market. Treatment and convenience became table stakes. As the standard of care improved, HIV/AIDS evolved from being a life-threatening disease to being a chronic disease. The market moved on to consider cure and prevention. This is an example of a market that has evolved dramatically over time.

CASE STUDY: THE ANTI-VAXXER MOVEMENT

Another example that is critical to consider for vaccine marketers that has left a mark on the healthcare landscape for vaccine brands is the "anti-vaxxer" movement. It started with a now-debunked article in the *Lancet*, by a now-discredited UK doctor, linking childhood vaccinations to increased rates of autism. Various celebrities fanned the fire and conspiracy theories spread in the following years. Social media played a role in spreading misinformation as well. This created fear and concern for some people. The result? In 2019, the United States saw the highest annual number of measles cases in more than twenty-five years.[7] A surge in rates of pertussis (whooping cough) was also recorded as a result of this cultural/social macroforce.[8]

Competitive Analysis

Sun Tzu's premise rings true—to beat the enemy you need to understand them, intimately. In fact, once you determine who your core competition is, you want to become them, step into their shoes, so that you can deeply understand them. Our industry lovingly calls this a competitive simulation. Make sure when you go into it that you don't stop at the obvious questions but understand what it feels like to be the competition today and anticipate where they are going in the near future.

This process begins with a complete and thorough analysis of your competitors. Define who your competitors are, what they are doing, how they are doing it, and how your customers perceive them. Once you establish the current state, predict and forecast their next moves: launches, indications, strategies, initiatives, new promotional campaigns, and so on. Finally, anticipate new competitors. These might not be in the marketplace yet, but they may have assets in development or waiting for regulatory approval. The marketplace is constantly evolving. Anticipating those changes are critical as you develop your brand plan.

Defining your competitive set is an important first step. Begin with the brands that are indicated for the same disease. Consider different modes of action, why they are used, and in which patients. Physicians will always explain that they select a medication due to its efficacy and safety. What they don't often acknowledge is that entrenched habits also play a big role in their prescribing behavior.

Depending on your disease state, some doctors prescribe medications off-label. This is fairly common in the oncology space. If this is the case, then you will need to include those brands as well. Finally, you might need to consider "work-arounds." Sometimes patients use other treatments, therapies, and solutions. These include non-medicinal interventions such as surgery, physical therapy, exercise, over-the-counter medications, and traditional Chinese medicine. You should consider these options as part of your competition.

CASE STUDY: DEPRESSION

An example of how this would look is depression. The depression market is defined by tricyclic antidepressants (TCAs), monoamine oxidase inhibitors (MAOIs), selective serotonin reuptake inhibitors (SSRIs), and serotonin and norepinephrine reuptake inhibitors (SNRIs). Sometimes physicians prescribe antianxiety medications to treat depression. Psychotherapy, cognitive behavioral therapy (CBT), interpersonal therapy (IPT), and electroconvulsive therapy (ECT) are often used to treat depression as well. Most recently, telepsychiatry has seen a surge in usage. Exercise, yoga, and meditation all help manage depression, too, and are indirect "competitors" that set the frame of reference. For antidepressants, perhaps sleep meds are one of your main competitors. Thinking through what method or medicine your brand would be replacing for that potential patient is critically important.

Evaluating Your Competitor

Evaluate and anticipate the strength of your competitor and what type of threat they pose. Sales performance, share of market, growth rates, and trends are key and great lag indicators. In addition, you need to look at net promoter score (NPS) and brand equity (BE) as indicators of your competitor brand's performance. You can get a sense as to whether your brand is heading in the right direction by looking at the scores over time.

You should also monitor your competitor's marketing and medical strategies. Do they have a large and established sales force? Do they have established relationships with KOLs? Do they have many clinical studies underway? Do they have robust scientific disclosures and publications? Are they at all of the key conferences? Do they have a bold social media presence? Do they have heavy television spend ($100 million to $250 million) with their direct-to-consumer (DTC) campaign or is it much more moderate and focused on digital, print, and other DTC vehicles ($10 million to $50 million)?

It is also important to evaluate not only the brand but the company as well. What is the competitor's financial status? Do they have more resources to put behind the brand or are they struggling to allocate resources to other brands they have in their pipeline? How vested are they in the disease state? Is the therapeutic area their core business, like Novo Nordisk and diabetes, or is it part of their portfolio, like AstraZeneca or Sanofi and diabetes? Are they an established company with a strong reputation that applies traditional marketing methods and closely follows guidelines and regulations? Or are they a new bio-tech launching their first and only brand, a company that might take more risks with a fresh and innovative approach?

A deep dive on your current and future competitive set is critical. You need to ask questions from many angles: business, branding, organizational, system, people, and others. Ask questions about all these key areas:

- **Their evidence/science.** How do they stack up with their evidence? How did they investigate their primary and secondary end points, inclusion/exclusion criteria, and so on? What do you think they will be able to claim promotionally? What exactly is in their package insert (PI), also known as a product label? Will that differ from what is already on the market? Will it command a price premium?

- **Their target.** Who is their target patient? What is their insight? What do they stand for, and how have they achieved that perception in the minds of consumers, or customers? How do they communicate it? Is it fractured among stakeholders with HCPs and patients, or is it connected? Does the payer also believe it?

- **Their organizational hierarchy.** How do they rank within their health and wellness company? Are they able to make decisions quickly or are they bogged down by hierarchy or perhaps by the bureaucracy of an alliance/partnership agreement? Is there evidence that they are a strong outside-in thinking organization? Are they able to command resources or are they the "ugly duckling" of their portfolio?

- **Their business strategy.** Is it hard to access their product? What has been their approach to access/pricing/reimbursement/health systems/payers? What is their approach to customer experience? Have they leveraged any tailwinds or mitigated any headwinds? Who do they perceive to be their main competitor?

- **Their initiatives/actions.** What key actions or initiatives have gotten them to where they are today? Do you think they are successful? What initiatives, tactics, or programs do they have in place and are recognized for? Have they taken any actions to steal share from their perceived competitors?

- **Their horizon.** Where are they in their life cycle—do they have additional data, or indications in the future? For the United States, do they have impending wins in the payer arena?

Understanding the competitive landscape is one of the hallmarks of a strong marketer. However, determining your direct competition versus indirect competition is taking the job to the next level. This involves understanding segmentation, which we discuss in chapter 2. Making a choice in terms of whom you need to win against and how you can displace them will help you determine the best path forward.

The tendency will be to include any and all competitors in your category. We would recommend focusing instead on your core competitors, the top two or three. If your target doctor or target patient is choosing a medicine or solution, which is the main product on their mind? Which is the one that your brand would replace in their mind, and why? That's the one you need to consider and focus on.

Write Your Competitor's Plan

A best practice that we have utilized is to actually write your competitor's brand plan. Ask a key brand associate to partner with someone on the market-research team and someone from the medical department. Ask them to take two weeks to write an abbreviated brand plan as if they were one of your main competitors. Once you have a few of these completed, distribute them to the broader team and then hold a half-day meeting in which each team presents one competitor's brand plan. Not only will you have a deeper understanding of your competitors, but you will also have a much better appreciation for the marketplace. You will be better able to anticipate your competitors' moves well in advance. You will be able to make better trade-offs with regard to your own brand plan decisions. As an added benefit, the team members will gain incredible experience from participating in a competitive simulation.

Key considerations when writing your competitor's plan include the following:

- **Company:** How healthy is the company? What does their profit and loss (P&L) statement look like? What does their pipeline look like? Do they have other assets, concerns, priorities? Are they innovative or conservative? Do they operate as a biotech start-up or big pharma? Do they regularly have deep investments in R&D and/or sales and marketing?

- **Capabilities:** Are they an industry leader? Are they a marketing powerhouse? Do they have deep DTC expertise? What is their digital expertise? How do they work with the FDA and regulators? Do they have the scientific expertise? What are their manufacturing and distribution capabilities? Do they have deep payer expertise?

- **Brand:** Are they innovative? Are they established in the category? What is their reputation? What do customers think about the brand? What is their BE? What is their NPS? What are the NILEX? What is the publication strategy? What is the social media and digital strategy? Are they big investors in sales and marketing? What is their promotional strategy?

CASE STUDY: PROZAC AND PAXIL

Some teams can become too focused on the competition and have "knee-jerk" reactions to their competition's strategy and tactics. A classic example of this is how the Prozac team reacted to the launch of a new competitor, Paxil.

Prozac was the market leader and was perceived by HCPs as being more of an activating antidepressant, useful for the patient who had lethargy. Paxil launched and claimed the anxious depressed patient. Paxil positioned itself for the depressed patient with anxious symptoms, a patient profile that fit perfectly with its clinical benefits and an FDA approval to treat social anxiety disorder (SAD). Importantly, this was a segment of the market that had not yet been developed and that no brand yet owned.

The Prozac team saw a decline in market share and responded by moving away from their core benefits and highlighting efficacy in anxious symptoms, chasing the competition. Prozac moved away from its "true north" in response to this competitive threat. In the customer's mind Prozac was well positioned as a strong and effective antidepressant for the lethargic depressed patient suffering from low mood and needing a boost—the stereotypical "depressed" patient. As it chased the competition, Prozac not only was unable to change its positioning in customers' minds but also started to lose some of its base. Its new positioning confused customers, as Prozac tried to be all things to all people. This loss of focus undermined its BE and eroded its market share. Eventually Prozac course-corrected and got back to its core target patient, but the damage to the brand perception was already done.

Let's Get Real: SWOT Analysis

Finally we consider tools such as the SWOT analysis. No marketing book would be complete without one. One of the first and major activities a brand team does is to put together a SWOT analysis.

The SWOT analysis is a look at your brand's strengths (S), weaknesses (W), opportunities (O), and threats (T). Opportunities and threats are external factors while strengths and weaknesses are internal factors. In the classic grid, the left-hand column (S and O) are things that grow your brand, and the right-hand column (W and T) are things that shrink or challenge it.

During brand plan reviews, we see the SWOT slide—it contains lots of details and interesting information. But there is no clear conclusion, no call to action, no aspect that translates easily to the brand. It is information, but information is only useful when it's been translated to decisions and action. Therefore, we aren't the biggest fans of this exercise, or at least the way most marketing teams do it. Instead of sitting down and writing the SWOT, we prefer to accomplish the same things but in a different manner: external pulse (step one), customer deep dive (step two), and brand assessment (step three). By the time you finish these steps you will have identified all of the SWOT inputs and a practical way of applying them to the choices you are making for your brand's success.

Of course, some organizations are beholden to SWOTs. If you are working for one of those organizations, make the most of this exercise. Seeing through opportunities and threats enables a team to think about the future state and the environment. Sometimes it brings to the surface ideas that will future-proof your decisions for your brand plan. But think about whether the SWOT slide needs to be in the presentation and make sure you don't spend more time on it than it is worth. If your team insists, then develop it, take out one or two key nuggets, clearly state the implications, and move on.

Cross-Collaborate to Validate What Matters

Brand teams use various tools and processes to analyze the external environment. Many of these can be tested and validated in the real world with some basic market research. They can also be vetted by leveraging the expertise in your organization or by reaching out to a few trusted advisors and subject matter experts. Creating a cross-functional team to workshop together is a vital way of ensuring that your thinking is correct from the foundational stages. It is also a wonderful way to ensure buy-in and to make sure that you don't have people rejecting your ideas and thinking simply because they were not part of the process and/or feeling slighted because they were not included.

When deciding which trends and environmental factors warrant inclusion, think about the potential impact on the market and your brand or portfolio. If one of these factors were to come to fruition, would it have a seminal impact on the opportunity for your brand? This context will allow you to be prepared for and apprised of opportunities and barriers.

Modern Marketing

As you consider the environment your brand is entering, step back and think about the various pressures on the healthcare system. You need to consider policy, regulation, and industry pressures. The pressure to reduce costs is being felt by government, providers, payers, pharma, and patients. Access is another major consideration. Universal access continues to be an emotional and heated topic. As a result we are seeing disruption, consolidation, and innovation at a pace not seen before. This provides many challenges as well as opportunities across all sectors of healthcare.

Consider consolidations across various sectors, including both horizontal and vertical integration. The role of technology, data, and digital health has accelerated. Providers are reimagining care. Patients' expectations, knowledge, and access to health information are changing as well. The concept of "care anywhere" is rapidly taking hold. Pharmaceutical companies are rethinking their sales force and promotional models. Research and development is evolving as AI enters clinical trial development, proving that value takes hold in the earliest stages of development. COVID-19 has taken all of these trends that were just starting but didn't have much urgency in face of the inertia of a large and established industry and accelerated them tremendously.

Take telehealth, for example. In its nascent stages it faced the challenges of getting properly reimbursed, issues with cross-state licensing, and concerns that virtual is not as good as face-to-face. Once COVID-19 came, swift changes in legislation ensured that telehealth would be reimbursed at the same rates as an in-office doctor visit and that doctors providing telehealth services were allowed to work across state lines as licensing requirements were waived; in addition, the value of virtual care became apparent. According to a McKinsey report, the use of telehealth has increased from 11 percent of consumers pre-COVID-19 to 46 percent during COVID-19. The market is expected to grow from \$3 billion to as high as \$250 billion.[9]

When thinking about the landscape your brand will operate in, think through the care path and how the places of care are changing. How, when, and who influences the decisions being made? Also think about how expectations of diagnosis, treatment, and prevention are evolving. Look at technology's role in the current landscape and what impact emerging technology will have in the future of the disease your brand addresses.

Key Questions

- What are the key trends taking place in your disease state?

- How will those trends impact your patients, HCPs, payers, and brand?

- What are the directions of these trends? Are they emerging or waning? Do they represent headwinds or tailwinds for your brand?

- Who are your main competitors and what are they doing to drive their success? What is their superpower? How much equity do they have in the market today? Are there any perceived gaps in their strategy?

- Are there any new competitors on the horizon?

- Have you considered any indirect competitors?

2

CUSTOMER DEEP DIVE

Key Learning Points

Stakeholder Analysis: The *Who*

Segmentation and Targeting

Patient Journey: Moments of Truth

Insights: Finding the *Why*

BRAND PLAN Rx CHOICE MAP

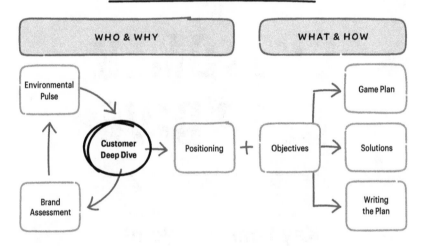

WHO & WHY

WHAT & HOW

Environmental Pulse

Customer Deep Dive

Positioning

Brand Assessment

Objectives

Game Plan

Solutions

Writing the Plan

THE PREVIOUS CHAPTER focused on the first step in the brand plan: understanding your environment. We now turn our attention to the second step: understanding your customer.

Before we begin, let's define "customer." As we mentioned in the introduction, one of the complexities of healthcare marketing is that there is no single customer—we market to doctors, nurses, payers, patients, caregivers of patients, and others. However, a note on terminology: We use the word "consumers" to refer to patients or their families. Consumers are still customers, but they are the end users of our product and are typically not HCPs. So "consumer" refers to the patient or their caregiver, while "customer" refers to all customers: patients, caregivers, HCPs, payers, and any other stakeholders.

Once you have a deep and solid understanding of your brand's environment, trends, and competition, it is time to move to the next phase of our outside-in thinking: understanding your customer. This involves understanding your customers deeply, feeling things from their perspective, walking in their shoes, spending a day in their life, seeing things through their eyes. Only when you obtain real empathy for your customer will you truly understand the market.

Uncovering their needs, wants, frustrations, and tensions better than any of your competitors is your responsibility as a marketer. This is known as customer insight. The importance of insight cannot be overstated. It is deep insight—understanding of your target customer— that separates astute marketers and meaningful brands. There is a

THE CUSTOMER INSIGHT PROCESS

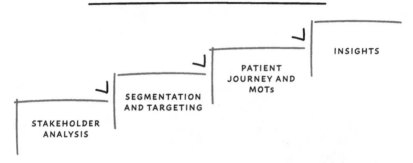

process to gathering it that is part science and part art. It requires a relentless and curious mindset. The process begins with stakeholder analysis and then proceeds to segmentation and targeting. Once you define your segment and have chosen your target patient, you then develop a patient journey. This helps you not only to understand what your consumer is experiencing but also to identify key moments of truth (MOTs) and understand their needs, frustrations, challenges, and opportunities. This sets you up for the final step, which is mining for key customer insights.

This is an iterative process. As you collect information, knowledge, and data, you build out the concepts and evolve your thinking. While we have introduced the concepts in a linear fashion, to obtain richness and depth it is best to continually revisit previous work as you progress, making tweaks and improvements until it all comes together.

Stakeholder Analysis: The *Who*

As you begin your insights discovery you need to consider this exercise through the lens of a number of stakeholders. In healthcare marketing the main stakeholders are the patient (consumer), the doctor (prescriber), and the insurer (payer). Depending on the disease state, you could also have the caregiver (also consumer), nurses, clinic staff, support groups, and health coaches, among others. For every disease, the patient and the doctor are without question the key stakeholders whose insights need to be considered. Which of the two is the key decision maker, and their relative influence, depends on a number of variables such as the disease itself, the healthcare system, culture, education, and income. Social determinants of health (SDOH) also play a key role and need to be incorporated in your analysis and thinking. The disease state plays a major part in determining the "weight" given to the influence of the patient or the doctor.

In oncology, for example, a cancer patient's insights are extremely important. However, as a result of a number of aspects such as complexity of care, innovative science, urgency of care, and emotional burdens, the oncologist's role is elevated even more. The oncologist becomes the key decision maker, and patients and families generally follow the treatment plan the oncologist outlines.

The other end of the spectrum would be erectile dysfunction (ED). Doctors rarely (if ever) bring up this topic during routine exams. Only if the patient mentions ED, and usually in a roundabout manner, does the doctor suggest a few options or courses of therapy. More often than not, the patient will build up the courage and ask their doctor for the medication by name (Cialis or Viagra), completely avoiding the awkward discussion regarding symptoms. There is an unspoken agreement. Nowadays, with the introduction of smartphone apps dedicated to men's health, little to no human interaction is even required for an ED prescription. This takes the prescribing doctor almost completely out of the decision-making process and makes the patient behave more like a "true" consumer. This concept of relative importance of customer types comes up again when you are determining your brand's positioning, a subject we address in further detail in chapter 4.

One of the first decisions to make is this: Are you mining for insights primarily from the perspective of the patient or the physician? The answer is both, but with some nuances to the methodology. Ideally you should collect insights and conduct market research with a sample of each of your key stakeholders. But the weight and relevance of each input depends on the disease state and should be customized for each brand. At the very least you are balancing between patients and HCPs, payers being the next most important.

Segmentation and Targeting

Segmentation is the process of separating the defined market into customer groups with similar characteristics, needs, attitudes, behaviors, and/or choice drivers. This yields a deeper understanding of the people who form the market. This is a landscape of "like" people grouped into segments, allowing for a choice of whom you wish your brand to target. Targeting is the process of deciding which segment you will focus on. Identify your primary segment, the target group of customers upon whom you will focus your energy, resources, and time. Then ensure your brand positioning, communications, promotions, and so on will resonate with this group the most. This does not mean that you will ignore the other segments or that your brand will not be used or needed by them. But it does mean that your brand absolutely has to resonate with the primary segment.

Let's Get Real: HCP Segmentation

Most pharma companies spend lots of time and money on physician segmentation. We have seen this done across numerous companies for countless brands in just about every disease state. But it isn't always the best use of your limited time and resources. HCP segmentation can be very useful when your brand's intended use is very clear at the outset—if so, it is the most effective way to build your communication plan. However, we believe that brand positioning should be anchored on patient insights. That doesn't mean that you should ignore HCP segmentation or insights. But if you need to make trade-offs (and we all do), then an in-depth attitudinal HCP segmentation solution might not be the way to go.

The weakness of attitudinal segmentation solutions is their reliance on typing tools that HCPs don't want to complete unless they are paid as market-research respondents. Most of the times we have completed HCP segmentations, we have learned that doctors fall into one of four segments (split roughly evenly): innovators, early adopters, traditionalists, or laggards. We have never seen brand teams successfully develop different messages for these different segments; rather, these segments are intended to be used in sales force implementation. But in reality, most sales force segmentation models are driven by prescribing behavior or potential. Perhaps this is a provocative statement, but we believe you could throw away most HCP attitudinal segmentation solutions for the purpose of brand development and it would not change the trajectory of your brand equity or sales uptake.

Instead, we think it is more important to spend some time understanding how HCP types relevant to your therapeutic area think and behave. For example, if you are launching a pain medication, then internists, pain specialists, rheumatologists, neurologists, and orthopedic surgeons might all be of interest. Each specialist is unique and has chosen that specialty for a reason. Understanding why someone became a neurologist is more helpful than segmenting neurologists to find that they roughly fall into one of four segments. Neurologists think, see, and approach treatment in a very different manner than pain specialists. Find out *why* and you might just get nearer to an insight.

Other types of HCP segmentation could then be used later when it comes to tactical implementation. For instance, segmentation that sorts HCPs in terms of how they seek or consume information would be a very useful tool for your omnichannel marketing plan. Patient segmentation, however, is very important and relevant for brand development. A deep understanding of your patient segmentation will determine your appropriate target, the patients' specific insights, the solutions they need, and how you communicate with them. You will use it when you complete your patient journey.

CASE STUDY: CYMBALTA

The depression treatment Cymbalta provides an excellent example of patient segmentation and targeting. A Harvard Business School case study on Cymbalta shows the impact this had on the medication's brand plan.[10] The depression market was highly competitive and already very segmented by the time Cymbalta was developed. The brand team began their exploration by conducting extensive market research that included qualitative and quantitative surveys designed to gain a deep understanding of patients' needs, wants, preferences, behaviors, and characteristics. From this, the team was able to identify seven unique, distinct, and identifiable groups of patients. They gave each group a psychographic handle, such as Hurting Helen, Functioning Fran, Complex Carl, and Anxious Anne, and developed a specific profile for each one. The profiles identified in detail the demographics, treatment history, current disease presentation, and other characteristics for each of the segments. The brand team was able to quantify and measure each of these elements.

In the Cymbalta example, segmentation research provided the marketing team with enough information and context to identify a key target segment. The most important variable was the information and data related to painful physical symptoms associated with depression. Hurting Helen was the patient segment that ideally matched the product profile of Cymbalta and represented an unmet need. However, Hurting Helen represented only 8 percent of the patient population. The brand feared getting niched and competing for only 8 percent of the depression market. They considered targeting other segments that were not an ideal match to the brand and were very satisfied but represented a larger share of the market. This is a classic dilemma: Do you select a target market for your brand that fits perfectly but is relatively small, or do you select a target that is larger but not an ideal match with your brand's attributes?

After much debate and some further analysis of the market research, the team learned that painful physical symptoms show up in the other patient segments as well. The decision was to focus on the 8 percent of patients who were Hurting Helens and whose main symptoms were the painful physical symptoms of depression. This turned out to be a brilliant decision as the messaging resonated perfectly with the Hurting Helen segment and had tremendous success and outcomes, as seen by the doctors and patients. The team also learned that patients in the other segments presented with painful physical symptoms of depression, but their doctors did not associate those symptoms with depression. This key insight led to the need for a market-conditioning campaign designed to help doctors recognize the painful physical symptoms and associate them with depression.

You can see a direct link from the segmentation and targeting work all the way down to the direct-to-consumer (DTC) campaign "Depression hurts. Cymbalta can help." The campaign depicted a patient who displayed all of the signs and symptoms of Hurting Helen. It informed people of the painful physical symptoms associated with depression and educated consumers on the link to depression and how Cymbalta could help. The catchy tagline nicely captures the sentiment: "Depression hurts, but you don't have to."

More importantly, as a result of the perfect fit and tremendous success with the Hurting Helen segment, the brand was able to credibly move to other appropriate patient segments, such as Functioning Fran and Complex Carl. Doctors saw the brand's success with Hurting Helen and, with the help of the market-conditioning campaign, started putting other patients with painful physical symptoms on Cymbalta.

Nike is another great example of the concept of being very specific to a target customer and benefiting from a halo effect. The target Nike consumer is the "serious athlete," yet many "wannabe" athletes and "weekend warriors" buy plenty of Nike shoes and athletic wear.

Watch Out

Brand teams are often excited when they get the results of a segmentation solution and have to pick a target segment. A common mistake is failing to align on the selection criteria with the cross-functional team before seeing the results. You then end up with people having a "feeling" about Segment B or a "hunch" about Segment D. This can all be prevented if the selection criteria are prioritized and weighted in advance.

Another common mistake is not making a clear choice: We like Segment B, but we think Segment D should be a secondary target and Segment A should be our tertiary target. This results in fielding market research and looking for results that might not "wow" Segment B but are acceptable to Segments D and A. This is the guaranteed path to mediocre marketing. Be bold and make a choice; anchor on your primary target and go all in! It is much better to resonate strongly with your key target than to appeal mildly to a wide array of patients. Without a key target patient your brand will mean nothing to everyone.

Modern Marketing

We believe conventional segmentation and targeting may become obsolete. In fact, learning about the *who* (customers) has recently evolved more than ever before. Thanks to the magic of data and analytics, we are able to deeply understand people at an individual level: their wants, needs, trust networks, fears, and desires, as well as whom they connect with, and how and when, in real time. While many still employ a quantitative segmentation analysis and then a selective targeting approach, other health and wellness marketers have started to understand microtargets through AI. Patterns and clustering of insights help them to identify a landscape of needs that are not being met. The beloved term "big data" is just starting to become actionable in the health and wellness sector. It is defined by extremely large data sets that may be analyzed computationally to reveal patterns, trends, and associations, especially relating to human behavior and interactions. Big data, as the essential element that powers up any AI, is set to be the future fuel of the world.

While segmentation isn't dead, it is evolving. In lieu of group identification personas (that is, groups with interesting psychographic names, such as Hurting Helen and Functioning Fran), we have 3-D pictures of microtargets whose current and future need states can be revealed and absorbed through real-time data rather than a one-time survey. In the health and wellness industry, the layers of who the person is and how they cope as a patient, whom they trust, and what it takes to drive decisions embody a complex, robust, and real-time understanding.

As technology has enabled more data, touchpoints, recency, frequency, and intersection points, we are able to learn more, with greater depth and breadth, in real time. The awareness that there are many types of people who could benefit from our product and a new ability to personalize messaging like never before warrants a new way of thinking about targeting. Many organizations are still in the crawling or early walking phase, trying new behavioral inputs of data to depict their customers. But, for the marketers who wish to go further, we would recommend exploring a customer-centric approach; the practice of personalization will ultimately benefit the outcome. If you employ the conventional approach, experiment with an additional data exploration to understand subtargets within the target. At the very least, your solutions will be brighter and more insightful and will likely give you a competitive advantage.

This form of hypertargeting exists in everyday marketing in other industries. Think of how Amazon knows what you are searching for—next time you log on to your device, you seem to receive advertisements for what you were shopping for. Think of how Netflix knows what you like to watch and recommends similar shows. This type of personalized segmentation has been slow to the healthcare space for a number of reasons. Understanding how a prescriber likes to consume information, what type of information they need, and when they need it are all techniques that are helping in the segmentation of the provider space. For example, if you have a physician who has searched for specific research about a disease state and has downloaded a certain type of research paper, then you know how better to meet the needs of that physician and can tailor your future communications.

Patient Journey: Moments of Truth

You have completed your segmentation exercise and selected a target patient. It's now time to create a patient journey map. A patient journey is the sequence of events your target patient experiences, from the very earliest stages to the end of the illness. It can begin as early as their awareness, concerns, worry, or just thinking that something is possibly wrong. It can end with cure, maintenance, relapse, or even death.

The patient journey is an internal tool that will

- help your cross-functional team achieve a unified understanding of the target patient experience,

- provide a view of the patient ecosystem,

- identify key moments that the patient experiences while managing their condition,

- contextualize and refine the important insights you have gathered, and

- uncover gaps in your customer understanding.

Hopefully by now your team has a repository of information, reports, research, sources, and some recurring themes pertaining to your customer findings. You also have one more source of information to add to it: your segmentation solution and the corresponding information about your target patient. Remember, your target patient is your "true north," so subsequent patient research should be based on this target.

Here are some key reminders when mapping out the patient journey:

1 The patient journey is always brand agnostic. The journey is for the target patient who is managing the condition/disease state of interest. This should not be the Brand X Patient Journey.

2 In keeping with the point above, you must take yourself out of your pharma/med device/healthcare company role while doing this exercise. Remove your bias for your company and your product. Forget your medical knowledge. You should channel your patient's thoughts and feelings to the best of your ability.

3 Use first-person statements when capturing patient insights, actions, and behaviors. And use your patient's language, not your language.

4 Capture not only what is happening functionally along the journey but also what is happening *emotionally*. What is the patient feeling, thinking, and worrying about?

Watch Out

Some people confuse the patient journey with the patient flow model (sometimes referred to as the healthcare transaction model, or HCTM). These two frameworks are not the same. The patient journey maps not just the actions but also the emotions the patient feels and the things they say (or leave unsaid) along the way. By contrast, the patient flow model, or HCTM, is a transactional model that quantifies and seeks to understand when patients seek diagnosis and treatment. It is a useful tool for forecasting but won't help you understand the experience through the lens of the patient. We discuss the patient journey and its usage in more detail in chapter 5.

Another limitation with the HCTM is that it starts with patients experiencing symptoms. However, we know from years of insight work that patients experiencing conditions with a strong familial or genetic link have been forming their views of that condition long before they have it. People living with type 2 diabetes, for example, frequently have a parent, grandparent, aunt, or uncle with that condition. They often have childhood memories of their relative injecting insulin, and that informs their patient journey. Life, disease, and treatment are never a straight line and decisions are full of worry, stress, frustration, and concern. For that reason, you should limit your reliance on the HCTM framework when creating a patient journey.

How to Map a Journey

Reviewing your existing research is a good place to start when developing a journey for your target. Did your target patient have any awareness of the condition before diagnosis? Was it already on their radar? What are the initial symptoms and what are the events

THE PATIENT AND PHYSICIAN JOURNEY: COVID-19

BOTH GROUPS FEEL WEIGHT OF ISOLATION AND HELPLESSNESS AND ARE READY TO GO HOME TO FRIENDS AND FAMILY

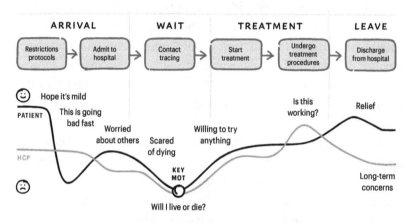

Modified from UNC Kenan-Flagler Business School MBA class project, Fall 2020

or triggers that are likely to make them more prominent? Are there coping mechanisms they use to stall for time or delay going to the doctor? What do they google? Most google searches consist of three words. What are the three words your target patient uses? Is there a defining moment that will prompt them to seek treatment or change treatments? How do they interact with their HCP? Do they even fill their first Rx? Their second one? Who do they talk to about their concerns, and when? If they experience side effects, what do they do?

While we do have some tools that help us to organize our thoughts, dig deeper, and learn more, this part of marketing is more art than science. It is where the nuances come in, words matter, and nonverbals play a big role—what is not said and done is often as important as what is. Many companies hire behavioral scientists, psychologists, and anthropologists to help them refine the insights and map the journey. We've even hired a monk to help our teams dig deep into the feelings, frustrations, and tensions of patients and providers. Be creative and get your information and analysis from anywhere you can.

The patient journey will have high points and low points. It should include the full range of emotional aspects of the journey: the points that are filled with worry, concern, frustration, tension, and confusion as well as those filled with hope, optimism, relief, and joy. Your journey needs to be developed from the perspective of the patient but should also include information about the moments when that journey intersects with the caregiver, doctor, nurse, pharmacist, insurer, advocacy group, and so on as appropriate. This might sound complicated, but here's how to keep it simple: the patient is the star of the show, the others are supporting actors.

Along this journey you will identify multiple MOTs. These are emotional revelations or physical events that trigger decisions or actions along the way. It's important to capture all of them if they are true to your target patient. You might end up with a patient journey with eight to twelve MOTs, but then what? We come back to simplicity and choices. It's essential that you pick two or three key MOTs for your team to focus on.

We recommend doing a quick quantitative assessment of the MOTs using a numeric ranking (1 to 7 will work). Look at three dynamics:

1   the level of patient tension or unmet need,

2   the ability of your brand to deliver on that unmet need, and

3   the impact on your business if you solve that unmet need.

Chapter 5 will help you better understand the third point as you seek to understand revenue drivers and sources of business.

Using that framework, we see in the example below that the team should prioritize MOTs 6, 7, and 3. MOT 5 is also a moment of high unmet need for the patient but one that the brand cannot reasonably solve and in which the company cannot logically invest. The beauty of this framework is that you apply the brand lens only after you have fully mapped out that brand-agnostic patient journey. It's essential to keep these two activities separate. If you mine for insights or craft a patient journey with your brand in mind you will never be truly patient-centric and will miss the *why*.

	MOT1	MOT2	MOT3	MOT4	MOT5	MOT6	MOT7	MOT8	MOT9
Level of patient tension or unmet need	6	3	6	2	3	6	2	3	4
Ability of brand to deliver on that unmet need	1	2	4	3	3	6	6	2	1
Impact on the business if you solve the unmet need	1	2	4	2	3	5	7	2	1
TOTAL	8	7	14	7	9	17	15	7	6

Now that you have your key MOTs, it is finally time to turn your attention to insights. Before we explore insights in depth we need to resolve a stakeholder issue. Back to the dilemma of whose insights: patients, providers, payers, and/or caregivers. The majority of brands will focus primarily on the patient, with some influence from HCPs.

Patient Insights

For all brands, it is best to look at the insights primarily from the patient's perspective, although in many critical disease states the patient is relying on the doctor to aid in their decision making. Someone with stage 4 pancreatic cancer, for example, will likely rely more on their HCP for therapeutic guidance than somebody who has gingivitis. But this is not to say that *consumer* insights are not important in situations when a patient's input is limited. The caregiver of the adult with advanced Alzheimer's or the parent of the infant with a rare disease will be a critical stakeholder, and their insights will likely be raw and visceral. However, for the vast majority of brands, the patient perspective should be the start of your strategic input.

HCP Insights

The HCP perspective is influenced by a number of variables. Take brands that require an injection or infusion. Often the prescribing HCP is not administering the medication or training, so their perspective is very different from that of the nurse who will administer the injection or the patient who will receive it. The best way to account for this is to conduct your research and analysis with two perspectives in mind:

1 HCP insights: this is straightforward—what the doctor thinks and feels about their role in treating patients. Their raw and direct perspective. Understanding the practical barriers and drivers of their decision making is also critical. For example, do they already have a lot of choices in this disease or condition? How difficult is the medicine to access?

2 Patient experience as seen through the eyes of the physician: this is when you ask the physician what *they* think the patient thinks and feels. This allows HCPs to step back for a moment, almost becoming a third-party observer with deep expertise, knowledge, and understanding. This will most likely become your most insightful and interesting perspective.

One of the biggest *aha* moments can come when you reconcile true patient insights (through the lens of the patient) with how the HCP believes the patient thinks and feels. There is sometimes an enormous disconnect that your brand or associated solutions can help address.

Insights: Finding the *Why*

When you think of an insight, pause and realize that you need to rely on empathy. Your opinion doesn't matter. All that matters is that you deeply understand what it feels like to live in the shoes of this patient, caregiver, doctor, or payer. In addition, it is vital that you help your colleagues to understand this concept as well. It is human nature to rely on your experiences and what you know to formulate your opinions. You and your team need to actively suspend that belief and try to get into the mindset of the customer. It's easier said than done.

An "insight" is the deep-seated need, fear, tension, frustration, wish, or desire that captures a not-yet-obvious driver of behavior, a deep inner truth from the customer's perspective. The insight unlocks an opportunity to inspire customer action and win in the market. Sometimes there's a tension, a conflict between the way the patient or doctor feels and the way they want to feel. That tension is heightened if the patient's and doctor's insights don't align and thus create friction. The insight highlights a gap between the current state and the desired state.

Uncovering insights can give your brand a key competitive advantage. The brand that truly understands its target customer will better meet their unmet needs and resonate much more strongly than competing brands, resulting in a better customer experience, stronger brand loyalty, and more brand equity. This is the essence of marketing. We all have brands that we prefer. Brands that you connect with emotionally, that just seem to "get you," that speak to you, that solve an unmet need, and that will get more of your business.

Nike is a classic example. Their "Find your greatness" campaign taps into the powerful insight that people want to push themselves to their potential, their limits. The everyday athlete strives to excel on their own terms. It's personal, your definition of greatness, defining a moment that helps you realize you can get to your goal and feel that inspiration and achievement. Everyone has an athlete inside of them.

By investing in yourself (and, of course, Nike) you can bring out the athlete in you.

To obtain these insights you must rely on multiple sources and always be collating and triangulating your data. Insights can come from primary market research such as quantitative surveys, qualitative interviews, in-home ethnographic studies, and online focus groups. They can also come from social listening, by tapping into Twitter and other social media platforms. You can also leverage your meetings with patient advocacy groups and HCP advisory boards. In fact, insights don't typically come from one source alone. You might get your initial "inkling" from one source and then use additional formats to validate your hypothesis.

Insights experts often rely on a deeper understanding of what people are feeling or what they believe but might not say. Words matter, but it's the meaning behind the words and the astute observation of what people aren't saying that generate a less common understanding of customer needs. This is especially true when it comes to health and wellness. Look at everything: the history of the category, how people have expressed themselves or not, and why. When you are doing market research, social listening, or meme assessment, look at words, heuristics, key body language, tone of voice, and codes. What are the common codes that your competitors haven't seen? Usually a customer won't share a deeper need overtly, so your assessment of the *why* is critical. Concepts such as secrecy, stigma, isolation, mind over matter, feeling drained, and soldiering on are some examples.

How to Mine for Insights

The tough thing about mining for insights is that you don't know what you're looking for until you find it. Be creative and curious as you learn about your customer. Simply asking questions and listening to the answers will likely not yield the deep understanding you need to get to the insight. There is so much more to pick up on. We have tried a number of creative techniques to dig deep and get to people's real inner emotions. The list includes therapy sessions and log books that require input at various times: before work when you are likely to be more optimistic or honest, during work when you're the busiest, at the end of work when you are dog-tired and more likely to be frustrated, over the weekends when you are more likely to be your real self.

As we mentioned, we have even hired a monk who could help us to read body language and subtleties, listen for silence, see group dynamics, and pick up on tone. One day while mining for insights at a market-research agency's plush offices, we realized that we were many steps removed from the reality of our patients' and doctors' daily lives: the busy clinics, chaotic home lives, stress caused by disease, and frustration with the healthcare system, among others. We were all highly paid, well-educated, and well-intended but biased employees trying to understand our market in a very "fake" environment. We needed to find a way to get a closer, more realistic sense of what our customers were truly experiencing.

Our monk was our answer. After a career as a religious leader in the United States, our monk had moved to Burma and spent seven years in intensive, silent meditation. This experience and training enabled him to see and hear things we didn't and to become a valuable partner in decoding customers' needs, tensions, and frustrations. In addition to teaching mindfulness, he has the ability to understand people and emotions to a degree that we were not able to. There were numerous times when we all drew a conclusion based on the information, data, and our observations, only to have our monk question our conclusion, provide a different perspective, and help us to see our customers in a much different and real light. This allowed us to truly see things through our customers' eyes and to deeply understand their unmet needs in a way that only a monk can.

CASE STUDY: DIABETES "SHRINES"

We have also visited patients in their homes or doctors after visiting hours, to spend some time in their shoes. When we spent time with people with diabetes, we learned that just about every single person has a "shrine" to the disease in their home. Especially those on injections. Some set them up in the bedroom, some in the kitchen, the bathroom, or a special place. They lined up everything that was needed and had a ritual they would go through to best manage their treatment. It showed us just how personal managing one's diabetes is and that it is a very private matter. For many patients, managing diabetes in public, especially around meals out at a restaurant, is very stressful and shameful. The "shrines" served as a symbol of patients wanting to keep their disease private, low-key, wanting to "mask" what was happening to them.

This insight gave us a sense of what this disease really means to these patients. In turn, this dramatically impacted the design of injectable devices and the tone of related promotions. We would never have found this out if we had only met customers in a doctor's office or a research facility. The deeper learning was that out in the real world, patients put on their best face and found the courage to manage injections and finger pricks multiple times a day, while at home, they went through a routine in a private place where they could muster up the energy to cope. Our marketing team worked on finding ways to make that ritual easier or to replace it entirely. If we could not replace the event, then how could we "fall into" the ritual rather than trying to disrupt it?

Understanding the *Why*

To identify insights, we rely on the well-known wisdom of Simon Sinek: "keep digging to the why, not the what." And don't forget that the *why* you land on has to be something actionable, something the brand can truly own that no other brand has capitalized upon. A big question that comes up often in health and wellness marketing is, Does that insight have to be something *only* our brand can answer? The answer is no. If you discover a deep *why* behind behavior that no one else has discovered and own it, then job well done! As long as you can credibly answer the need and rally behind it in all that you do, you can own it.

Understanding the *why* is the most essential exercise a marketer must perfect when building a brand. It is the key that unlocks the magic. If you get this right, the rest of the brand plan flows, your strategy resonates, your initiatives have meaning, your brand has relevance, and your customers (patients, doctors, caregivers, payers, etc.) become raving fans. Behavioral science shows that if we want to change behavior it is better to start at the high level—the *why*, before you get into the *what*.

Understanding the *why* shows your customers that you understand *them*. In return they will reward you with market share and brand loyalty. Miss the *why* and your brand is destined to underperform, as consumers will turn to other brands.

Understanding the *why* of your brand begins with understanding the wants and needs of your customer, the tension and problems that they are looking to solve. The unmet need that is holding them back. The frustration that they are feeling. Understanding begins by looking at the customer's wants, needs, tensions, and frustrations and then coming up with a solution, not the other way around.

In the healthcare space, no one *wants* your brand. Patients *need* it. They want to return to normalcy; they want things to be the way they used to be; they want to be healthy; they want to buy more time. But they don't want your brand. Rather they need your brand to achieve the things that they want. They need to see a doctor; get a prescription for medication; go to the pharmacy; pay for it; and take the right dose, at the right time, and in the right amount.

You have to ask *why* many more times than you would think to get to the real, underlying *why*. For just about any pharmaceutical product in the world, when you ask doctors *why* they don't choose your brand for more patients, the most common knee-jerk responses are "Because it is too expensive" or "Because I am comfortable with what I have used in the past." These are rarely the real reasons *why*. In fact, most people (doctors and patients alike) do not know how much medications cost. Only after peeling back the onion does one get beyond prices and finally arrive at the *why*. It is always driven by a key insight.

At the end of all the analysis, an insight is a fresh and penetrating truth about the target market and/or target customer that opens up opportunities to create enduring advantages for your brand.

Behind the *Why*

Health is intimate. It's personal. For many people, it's not easy to talk about how they feel about what they need, what they really care about, what they fear. So we need to go deeper. We need to understand why people feel this way. Go beyond words, listen carefully, observe actions and reactions, and look at how people behave—when they pause, the intonations in their voice. Most people (especially doctors) can't say what they really feel, or are holding back as a self-preservation tactic. Decision science—the study of why people fear or desire a change or are paralyzed or want more—is a nuanced approach to get to the *why*.

How do you know when you've gotten to the right insights? How do you know if you have mined deep enough, asked *why* enough times, reached to the heart of the matter? This is the question we get most often when we're building a brand plan. Align on the wrong insight and your entire plan will be off.

The answer is, it is very situational and context dependent. Every case is different, so there is no easy answer. What we can tell you is that when you *have* dug deep enough and found it, it is a eureka moment. You know it when you see it. An insight is a lot like love—it's hard to define, but once it happens, you know it. The alternative is true as well. If you aren't having that eureka moment or "feeling butter-flies," then you haven't found it. Once you do find it, it becomes fairly obvious. Your eureka moment is quickly followed by a *duh* moment. It all seems very obvious once you have your key insight. Your first thought will be, *Duh*, why didn't I think of that earlier?

In summary, be empathetic from your customer's point of view, remain relentless in your curiosity, and persistently ask *why*.

CASE STUDY: CIALIS

Cialis in the erectile dysfunction space is a great example of how to mine for insights in the pharma world. It is a classic David and Goliath story. Viagra, one of the largest and most successful brands in the world, was the prize asset of Pfizer, a large, successful, and well-resourced company that had singlehandedly created the ED market. They had a dominant share of the market, great brand equity, and outstanding customer loyalty with doctors and patients. They were spending huge amounts of money on advertising, both on TV and in print, and had a large and successful sales force. Brand recognition and sales were achieving record levels.

Eli Lilly, in contrast, was a midsize pharma company that had no expertise in the ED market or with this customer group; their expertise was in neuroscience and diabetes. Their resources were also very constrained—there was no way they could compete dollar for dollar with Pfizer. They had no equity in the market to leverage. It didn't seem like they had much of a chance as they were getting ready to launch Cialis in 2004.

As the marketing team got to work developing the brand plan, they quickly learned about the environment, the customers, and their brand. The perceived clinical effects of Cialis, based on initial feedback from stakeholders, seemed like a drawback: slower onset and longer duration of action (thirty-six hours; in contrast, Viagra worked fast and lasted for only four hours). Viagra also had the perception of being powerful and effective, exactly the equity you want in the ED market.

However, once the brand took a step back, looked at ED more closely, and examined the patient journey, they uncovered a unique insight. ED isn't just one person's disease; it impacts men's partners as well. Deeper conversations with couples revealed that they loathed the idea of having to be ready to "get on with it" as soon as the man was ready. The key learning came during interviews with the partners. They said, "It felt like there were three of us in the bedroom: him, me, and Viagra. Once he took the pill the clock started, and we knew we only had a few hours."

That was the clue that led to an ever-deepening understanding of the frustrations that partners in such relationships were grappling with. The presumed disadvantage of Cialis was actually an advantage, and the short duration of Viagra (a window of opportunity that only lasts four hours, not thirty-six) became a vulnerability.

The insight is that spontaneity is key to true intimacy. Couples wanted to regain their relationship by having intimacy, not only by having sex. And for most partners, the weekend or a weekend getaway is the time when they are most likely to relax, enjoy their relationship, and be intimate. Cialis gave couples the freedom to let the moment happen, versus the pressure of having sex within a four-hour window after taking Viagra. Interestingly, in France, Cialis was colloquially referred to as *le weekender*. Viagra was marketing to men and selling erections. Cialis was marketing to couples and selling restored intimacy in relationships. What at first glance looked like a satisfied market with little opportunity suddenly became attractive with this insight, which uncovered a huge unmet need.

The insight was brought to life in implementation of the brand's messages, promotional material, and DTC campaigns. Through this you can very clearly see not only the differences in the two brands but also how each focused on its core insight. Viagra's communications focused on the end consumer, the male, depicted in a very macho and aggressive manner—on the prowl, if you will (often depicted in a bar or a bedroom scene). Any females included in the advertisements were young, beautiful, and dressed and behaving in a very sexual manner. Conversely, the Cialis promotions introduced one of the most recognizable and famous pharmaceutical campaigns ever: two bathtubs overlooking a natural landscape. It featured a couple, much older than the main Viagra characters. The music was romantic, the pace of the commercial slow, the voiceover soothing. The couple were doing enjoyable activities that couples do. The clear focus was on the relationship and not on the act of having sex. This became known as the "When the moment is right" campaign.

The brand team built a brand plan based on a key insight that was meaningful and relevant to customers and resulted in a clear call to action for the target market. Eli Lilly was able to gain a leadership share of market (SOM) position across the globe without having to match Pfizer's spend dollar for dollar.

Cialis has since grown its worldwide revenues to more than $20 billion, overtaking Viagra as market leader in 2013. By 2015 it had left its main competitor behind, recording annual sales of more than $2.3 billion. It seems size isn't everything...[11]

Observation to Insight

The art of mining for insights is turning what might seem like every-day observations into key learnings. Some introspection is required, along with a deep understanding, relentless questioning, and abundant curiosity. Topamax, Aciphex, and Requip provide three excellent examples of the difference between an observation and an insight.

Topamax

Observation: Migraine sufferers want to reduce the frequency and severity of attacks.

Insight: It's not an attack. It's a condition.

Aciphex

Observation: Reflux patients want symptom relief that is fast and effective.

Insight: The symptoms that no one talks about are the ones that bother reflux sufferers the most.

Requip

Observation: Poor sleep plagues people as they get older and restless legs syndrome (RLS) is often the major complaint.

Insight: There's more to "restless" than poor sleep.

Insights across Cultures

One of the teams we worked with was charged with launching a medication in Japan for the treatment of pain associated with diabetic neuropathy. Global prevalence rates at the time suggested that 10 to 20 percent of people with diabetes experienced some form of painful neuropathy. Several epidemiological studies assessed diabetic peripheral neuropathy among US adult patients with diabetes and reported prevalence rates of 26 to 47 percent.[12]

However, the team in Japan was telling us something different: their HCP market research was indicating that the problem was almost nonexistent. Could the experience of people with diabetes be so different in Japan?

The team only had some qualitative research with HCPs at this point, and the customers were almost unanimously saying that their patients with diabetes did not complain of any burning, stabbing, shooting pain in their feet (the most common symptoms). They simply did not see this problem. It was clear that we needed to mine for more insights.

Two sources helped with next steps: a published literature scan and a quantitative survey of patients with diabetes. The literature scan uncovered an article that examined differing attitudes toward pain between Japanese and Euro-American subjects. Pain behaviors (i.e., expressions of pain) were less acceptable in the Japanese population. Because of this, the team felt that an anonymous, quantitative survey with patients would help shed light on the situation in Japan. The results were both reassuring and shocking. Reassuring in that about 15 percent of the respondents were experiencing some form of pain in their feet, in line with the global prevalence. But shocking because the doctors who had been interviewed were clearly not aware of this.

The question *why* hung in the air. The team followed up with qualitative interviews with people with diabetes who had symptoms of diabetic neuropathic pain. It took hours of interviews and skillful questioning by the moderator to "peel back the onion." Here are some themes that emerged:

- Some patients didn't realize the pain was associated with their diabetes and therefore didn't feel the need to mention it to their endocrinologist.

- Some patients suspected the pain was related to their diabetes and felt it was a sign they hadn't been managing their health properly. They therefore felt guilty and feared looking like a failure if they mentioned it to their doctor.

- Many patients felt guilty mentioning their pain to the doctor because there were so many other people in the waiting room. This reason was commonly cited and unique to Japan. Most patients don't have a specific appointment; they just show up on an assigned day and have to wait with many others. They typically get five minutes with their doctor to review lab results and talk about their medication. They felt it would be selfish and disrespectful to the other patients waiting to take more of the doctor's time by bringing up additional "problems."

The team also did a deep dive into the exact words the patients used to describe their symptoms. This was also eye-opening and uncovered another disconnect. One of the most common symptoms patients referenced was じんじん, or *jin jin*. This is hard to define in English, but essentially it's a tingling, throbbing numbness that might or might not be painful. When you are marketing a drug for pain and your target patient is most commonly citing じんじん as a symptom, things get tricky. So the insight was that pain is not binary, or there is pain behind pain—the ongoing discomfort is as debilitating as the typical pain symptoms doctors presume.

However, by constantly digging, using a variety of sources, and always asking *why*, the team was able to uncover a wealth of patient and HCP insights. This changed significantly the *what* and the *how* that the brand communicated to patients and HCPs, resolving an unmet need that neither party was aware of and essentially serving as a bridge between patient and HCP. The brand was wildly successful as it solved a problem for both the patient and the HCP.

Insights Tracker

The process of uncovering or "mining for" insights is continuous, evolving as the market and customers develop and change. It's represented in just one section of our process map, but the expert marketer is always looking for clues that will lead to an insight. Great teams have an "insights tracker" or something similar—a record of findings that are distilling and bubbling, along with their sources, to be tapped into as they begin or modify their patient journey, their HCP messaging, or their brand plan.

Force yourself to continue to observe, learn, empathize, and challenge all you know about your customer. Through keeping an ongoing repository and understanding the implications, this will become an instituted behavior and bolster your brand's success.

One way of keeping track of these insights is by listing them in a table. For each insight, note the customer type, persona, source, observation, insight, qualitative findings, quantitative findings, implication, and planned follow-up.

Modern Marketing: Co-Patients

Examine how the disease or prevention needs are affecting not only the core people but also anyone else in the patient's circle of influence. Some diseases impact family members or trusted loved ones as much as the patient themselves. You may need to market to the co-patients, not just the patient. Consider if the disease or wellness treatment affects another person in the target's universe. In a disease like Alzheimer's this is fairly obvious. Figuring out how family and friends are implicated in less obvious cases is more challenging. For example, in diabetes or ulcerative colitis treatment, there may be important aspects that haven't been explored or considered within a household or a close relationship ecosystem. Examine these and learn more about the impact of the disease, not only on the sufferer but on their closest connections—the *co-patients*.

Insights Summary

In this chapter we have discussed insights in depth via a number of tools, techniques, and examples. After working through all this it is helpful to step back and take a high-level perspective of the methodology. This will help give you some context in order to identify and align on how best to determine your brand's key insight.

Organize your analysis and thinking by first looking at your key stakeholders. Then focus on the target patient that you identified in your segmentation and targeting. Your prioritized MOTs from the patient journey you created will help your team understand the patient's true unmet need. Finally, consider all of the research and information you obtained while mining for the insights. This process will lead you to a single key insight that is meaningful and relevant.

Once you have completed the brand planning covered in this chapter, you've done some very heavy lifting. If you are developing a new brand, the customer deep dive will help set the foundation for your positioning as well as your brand plan. If you are working on an existing (on-the-market) brand, you are testing your current assumptions around stakeholder analysis, refreshing your insights on your target patient, and walking through their journey again with fresh eyes. Now it's time to do a deep dive into the product that you are responsible for—it's time to assess the brand.

Key Questions

- Who are the key stakeholders for your disease state/therapeutic area?

- Who is your target patient? Describe them psychologically, physically, socially.

- How does your target patient primarily define themselves? What are their goals in life? Their hobbies? What are they most proud of? What are their biggest regrets? What do they value most? How do they make their decisions in life? It's important that you understand them first as a human being before thinking of them as a patient.

- How does your patient approach health and wellness? How do they view their disease or illness? How do they characterize their relationship with their HCP? What is their understanding of treatment options? How do the patient and HCP each define treatment success? Are they on the same page?

- How does the HCP perceive your target patient and associated insight?

- For the HCP, what does it feel like to treat the disease? Does your target patient represent a small or large part of their practice? Is it rewarding or frustrating to treat this condition?

- For the payer, does this therapeutic area represent a large part of their budget? If so, it will be under close scrutiny. What are the therapeutic equivalents or other standards of care they will be using as a reference price? What do they see as the biggest unmet need in this area? What type of innovation will they need to see in order to grant a price premium?

- What does the disease feel like? What does it feel like to treat the disease? How do patients and doctors feel about what they have accomplished? What do patients (or doctors) wish for?

- What is your target patient's key insight?

3

BRAND ASSESSMENT

Key Learning Points

Interrogating the Brand

Features, Advantages, and Benefits

Differentiation Grid

Superpower

BRAND PLAN Rx CHOICE MAP

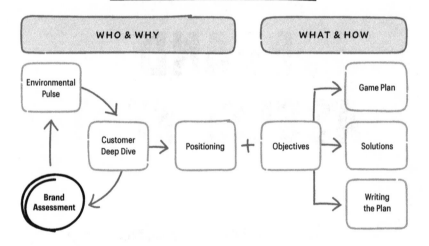

U P TO THIS point in our brand plan we have taken an outside-in approach by looking at the environment and the customer. It is now time to flip that thinking and take an inside-out approach. We do this by taking a deep, honest, and thorough review of our brand. Consider this an interrogation of the asset.

First we look at our brand's features, advantages, and benefits. Then we apply that to our differentiation grid. Finally, we determine the brand's "superpower." The objective of this section of the brand plan is to have an accurate, realistic, and aligned understanding of your brand and how it will compete in the marketplace. You might uncover some things you didn't know and some gaps that exist, or simply confirm what you already know about your asset.

Interrogating the Brand

For a prelaunch brand, this is when you get to "geek out" over all of the data you have available to you. You'll want to become a student of the asset and learn everything you can about the brand, service, procedure, and/or offering. In pharma this would include clinical studies and the molecule: for example, the inclusion and exclusion criteria, the primary and secondary end points, the dosing, the adverse events, the mechanism of action, and the device (if applicable). You are looking for any and every point of differentiation from existing and pending treatment options.

For on-the-market brands, this is a good time to understand the "state of your brand." Take an objective evaluation of your brand's performance. How are sales versus plan? What is your share of market (SOM) versus plan? What is your growth in dollars and volume versus plan? How has your spend been versus your budget? How have your brand solutions performed in the marketplace? Do you have any market research that will help you understand how your brand is performing in the marketplace? What is working well and what is not? Brand equity audits usually provide good data to summarize and share. Your "state of your brand" should be a snapshot of your brand's performance.

An additional aspect you need to consider is a full interrogation of your competition. We visited this in chapter 1 when we discussed the competitive analysis. This is a good time to review that work and ensure that you have done a detailed analysis of your key competitors. A deep dive into their labels, studies, data, publications, and promotional claims is a worthwhile exercise.

Your R&D and medical affairs team will have likely already highlighted some points of differentiation, but they are looking at it primarily from the standpoint of clinicians and might not see some value drivers from the perspective of the patient or payer. Don't underestimate the value of this exercise. It's critical to get a truly cross-functional perspective on this. We recommend a team that includes marketing, sales leadership, medical affairs, medical development, health economics, payer marketing, regulatory affairs, licensing and acquisition, patient advocacy, and public or corporate affairs. If you're on a global team, it's also important to get participation from a few of your key markets.

Watch Out

As you dig through the data and learn more about the brand and the science behind it, you are likely to find many points of differentiation or possible competitive advantages. You will identify many attributes that you think are excellent. Before you have even finished uttering your thoughts you may be drenched in a chorus of negativity raining down upon you:

- "We don't have data to substantiate that."
- "You'll never be able to make that claim."
- "That was not a primary end point of the study."
- "The exact mechanism of action is unknown."

Your research and developmental partners have been living the science, the protocols, and the trials. First, go to them to really understand the science. Why is our molecule different from a competitor's? Does its mechanism yield a new algorithm or way to approach the disease? Then push beyond the science and understand how the brand is answering what a person needs. Think through the context and the human value that is answered by the scientific value. It is important to put some guardrails in place when you begin this exercise. *This is not being done to generate promotional claims!* The purpose is to get out of your industry hat and put on your customer hat (HCP, patient, payer) and truly understand every aspect of the asset and how it differs from what is currently on the market.

For an existing brand, your brand assessment will rely more on numerous tools, including brand equity audits, consumer trackers, social listening, voice of customer (VOC) surveys, new scientific disclosures, and, of course, sales data. It's important to check that your customers' perceptions match your desired positioning. You might have great sales this quarter, but that could be due to something that is entirely off-brand. You'll find this out by checking your brand equity (BE) reports or VOC surveys. This might be fine in the short term, but if the reason your customers are selecting your brand is not unique or ownable, it presents a weakness and likely a departure from your strategy. It should be an immediate warning to your team to assess your campaign and messaging across your markets to check for alignment to your positioning, to ensure a sustainable competitive advantage.

Features, Advantages, and Benefits

You've now "interrogated the brand" based on your deep dive of the prelaunch data or the in-market information available to you. It's now time to translate the attributes you have identified into benefits that will be meaningful to your customers.

When assessing your brand features, advantages, and benefits (FAB), ask yourself questions related to the following categories:

- **Product features:** What are the specific attributes of your brand that meet a customer need the competition cannot or has not addressed? These are differentiating characteristics supported by data and evidence.

- **Advantage:** What is the functional benefit? What does the product actually do for the consumer/customer? What is the unique value? These are the consequences of the features to the consumer.

- **Emotional benefit:** When moving to the next level, what dominant feeling is generated as a result of using or recommending the product? This is the result of the benefits—what it helps your consumer feel, think, do, and become.

You have a great foundation for this now that you have your target patient insights and you know a lot about your key HCPs and payers. Let's take a look at a few examples that will help bring these concepts to life.

Topamax

Feature: reduces frequency of migraine attacks
Advantage: fewer days with migraines
Benefit: prevents migraines from hijacking your life

Cialis

Feature: has thirty-six-hour efficacy
Advantage: grants longer time frame in which to initiate sex
Benefit: restores intimacy—when the moment is right

Nexium

Feature: utilizes superior healing data
Advantage: not only treats reflux symptoms but heals the damage
Benefit: fixes the underlying problem

We will go into the FAB concept in much more detail in the next chapter, when we discuss the benefits brand ladder. We wanted to introduce the concept here so you can complete the differentiation grid. As mentioned earlier, this first half of the Brand Plan Rx Choice Map is iterative, so circling back is a good thing.

Differentiation Grid

Now you need to pressure test your thinking by using the differentiation grid to identify which benefits to focus on. The grid will help you determine which benefits will give you an advantage over the competition relative to the attribute's importance in the marketplace.

The simple matrix plots your brand's attributes versus those of your competitor along the X axis. Attributes that your competition is better in belong in the first column; attributes that are the same go in the

BRAND DIFFERENTIATION GRID

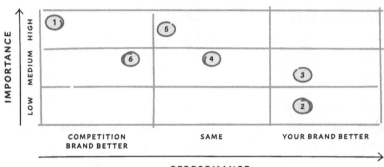

second column; and attributes that your brand is better in belong in the third column. Along the Y axis you plot the relative importance of these attributes to your customers: low, medium, or high.

Let's look at an example. Plotted on the matrix above are some key drivers of HCP preference for a brand that we helped launch:

1   **Onset of action:** This product works quickly.

2   **Dosing:** This product is easy to administer and has flexible dosing.

3   **Side effects:** This product has minimal side effects.

4   **Efficacy:** This product reduces symptoms and has robust results.

5   **Safety profile:** This product is proven safe.

6   **Cost:** This product is affordable for patients.

In this example, if we take a look at attribute 1, onset of action, we can determine that this attribute is very important in treating this disease. However, your brand is not as good as the competition. In fact, it is worse; it takes longer to resolve symptoms. To ensure the success of your brand, this is something you will clearly need to manage in the marketplace. Expectations will need to be set and you will need to come up with ways to help HCPs and patients manage until the efficacy kicks in.

If you look at attribute 2, dosing, the placement on the matrix indicates that your brand works much better than any other in the marketplace. However, the relative importance to HCPs is very low. This would indicate that a meaningful strategy might be to see how you can elevate the importance of this attribute in your HCP's mind. Perhaps the HCP doesn't perceive dosing to be an issue, while in fact it is a tremendous problem for patients—but the HCP doesn't see that because the complaints reach the nursing staff and never make it past them. Your brand strategy might be to find a way to bring that awareness to HCPs and hence elevate the attribute's importance, leveraging a clear competitive advantage of your brand.

This exercise gives you a sense of how to use the differentiation grid to determine which attributes and benefits you will focus on and how. It will also help make sure you are not launching or promoting a "me-too" brand. If you have picked an attribute that is already owned and satisfied in the market, you will need to carefully consider next steps. You also need to determine if your point of differentiation goes against current marketplace expectations. Leverage your market and customer insights to understand the key drivers of preference.

This exercise also helps highlight the potential strengths and weaknesses of your brand (another reason why we feel you can be marketing rebels and abandon the sacred SWOT slide, if your organization allows this).

In the grid above, we focus on highlighting the unique strengths of your brand. It is entirely possible that during the course of your brand interrogation and differentiation discussions you will uncover a significant weakness. How should you address it? If it involves patient safety or tolerability, it should be addressed head-on. Transparency is key, and guidance to HCPs and patients on how to mitigate the concern will set your brand up for success.

CASE STUDY: CYMBALTA

Let's turn again to Cymbalta for an example. One of the most common adverse events when starting Cymbalta is nausea. It would have been tempting at launch to push for simplicity of dosing: 60 mg as both the starting dose and the target dose (all the competitors' starting doses were the target dose, so no titration required). However, the best patient experience was obtained when patients were started at 30 mg for one week and then titrated up to 60 mg. This made the dosing less simple but minimized the nausea concern and set patients and HCPs up for a good initial experience with the brand.

This feature—the need to titrate—was a disadvantage. The differentiation grid helped identify dosing and titration as a specific vulnerability for the brand versus the competition. However, it was low on the differentiation level and low on the importance level as well; the team could see that other attributes were more important to patients and HCPs. This systematic process helped to highlight to the brand team that we needed to address this issue in the context of the competition and other brand elements. The team focused on those and mitigated any concerns with dosing and titration.

What Would You Say in a Café, or Write as a Newspaper Headline?

After completing your brand interrogation, your attributes and benefits, and your differentiation grid, you should feel confident in your brand assessment. But, as we'll continue to do throughout this book, you should challenge yourself and your team to strive for simplicity. An effective way to do this is by reviewing the outputs of those exercises and challenging yourself to write a newspaper headline that captures what you would want people to say about your brand in the coming year or two. Or by thinking about what a field representative might talk about after a sales meeting when they are debriefing casually. (What would stand out? What would they be excited about?) We are limiting you to a headline because it will force you to make tough choices: a brand that is trying to be all things to all people is one that is destined to fail. This is such a critical point because who our brand is and what its superpower is should be the backbone of how we approach our brand plan. The objective, strategies, and initiatives all need to serve the brand and how the brand is answering a need for our stakeholders.

Superpower

Just like each action hero in Marvel Comics' Justice League, every brand has what we call a "superpower." Each character has a unique ability that they use to help save the world from evil, just as your brand has one unique characteristic that it provides to patients to ward off disease. Your job is to find this superpower, unearth it, and bring it to life.

Just like our action heroes, you need to stick with the main superpower of your brand—the *one* aspect that is unique, clear, convincing, and meaningful, and that delivers results. When you try to become too many things to too many people, then you dilute yourself and become nothing to all people. It is also important to know the superpowers of your competitors. This is helpful as you plan your strategy and to make sure that your superpowers don't overlap. If they are the same, is yours convincingly better? If your superpower is the ability to fly, you'd better be able to fly better than Superman. But it's better if none of the superheroes' powers overlap. It works best that way. You want to home in on your brand's superpower that no one else has or owns.

The Z-Pak and Crestor are a couple of examples of a brand's superpower.

CASE STUDY: Z-PAK

Pfizer gained market leadership in the antibiotic space by launching Zithromax in a simple pack that allowed patients to complete their course of antibiotics in five doses. The "Z-Pak" was up against a more entrenched and effective antibiotic, Biaxin. This should have been a tough battle given the compelling data that showed superior efficacy in Biaxin, but Biaxin required twenty doses—a tough proposition for parents trying to get their unwilling child to complete a course of antibiotics.

The Z-Pak that Pfizer introduced came in an easy-to-use cardboard box that flipped open and provided detailed and clear dosing instructions. In addition to being unique it was also easy to carry. As a result of the packaging design, Pfizer was able to include some communication to patients regarding the cost, the importance of completing the full regimen, competitive claims, and further instructions—none of which is possible with the traditional pill bottle. It also helped you keep track of your doses; as you used one dose you could easily see the open spot in the package. It provided one more key advantage. As patients would get a pill bottle with only six Zithromax pills in it, it felt fairly empty; the pills rattled around in the big bottle. Patient research indicated that they did not feel like they were getting their money's worth. They were more accustomed to a pill bottle that was mostly full rather than mostly empty. The Z-Pak resolved the impression that Zithromax patients weren't "getting their money's worth." Its superpower of simplicity of dosing was too convincing and it quickly dominated the market.[13]

CASE STUDY: CRESTOR

Prior to the launch of Crestor, most statins focused on their ability to lower total cholesterol. Lipitor and Zocor dominated the growing market. AstraZeneca launched Crestor with a message that differentiated "good" cholesterol (HDL) from "bad" cholesterol (LDL). Crestor's early messaging touted the brand's ability to increase good cholesterol while lowering bad cholesterol. In fact, Crestor reduced LDL by 55 percent, while Lipitor reduced it by 43 percent and Zocor by 38 percent.[14] This messaging evolved to a singular focus and one superpower: fighting bad cholesterol. The promotional campaign utilized very clear and distinct arrows pointing down and stating that the medication lowers bad cholesterol, and by how much. As the first to market with this data, AstraZeneca did not need a super clever campaign—they could be simple and clear about the benefit they were providing.

Key Questions

- Are the benefits you have identified for your brand emotional? Are they the emotional result of the features and advantages of your brand? What does your brand help customers think, feel, and do?

- How is your brand different from what is on the market? Do customers value the functional and emotional benefits that it offers? Is your brand's differentiator contributing to brand preference? Or do you need to conduct market conditioning to make that differentiator more relevant? Have you looked at the clinical, patient, and economic benefits as a whole?

- What is your brand's superpower? Does it align to your customer's key insight? Does it solve their needs, frustrations, and tensions?

- Have you identified gaps in evidence to support your brand's superpower? What is the plan to obtain this data/support?

4

POSITIONING

Key Learning Points

Brand Opportunity Distilled

Building Blocks

Needs/Has Venn

Benefits Brand Ladder

Positioning Template

Positioning Testing

BRAND PLAN Rx CHOICE MAP

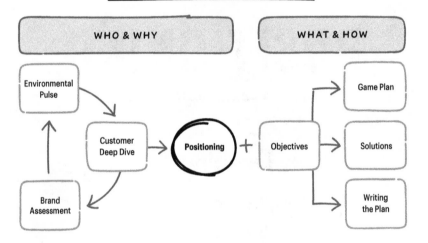

WHO & WHY

WHAT & HOW

Environmental Pulse

Customer Deep Dive

Brand Assessment

Positioning

Objectives

Game Plan

Solutions

Writing the Plan

P OSITIONING IS NOT something that is done annually as part of the brand planning process. It is the "true north" of your brand against which your plan should be validated and gut checked. It is a workstream in and of itself. For that reason, we'll provide just a basic overview of positioning. We could write a whole book on the subject but will only devote a chapter to it here because this book is focused on brand planning, not brand development.

Positioning stakes out a place for the brand in the heart and mind of the customer. It identifies the emotional and functional need your brand is fulfilling in the customer's mind. Positioning is an internal document that articulates the association you plan to build between the brand and the customer: a functional, emotional, conscious, and unconscious association.

Let's recall our discussion in chapter 2 of customers. Customers are anyone who might purchase our brand: HCPs, patients, pharmacists, payers (including government bodies). Included in that definition is a specific type of customer: consumers. Consumers are typically the patient, the end user of our brand who uses, ingests, injects, or applies our product or service. However, the consumer can also be the caregiver or decision maker for the patient—the parent of a child with cystic fibrosis, or the adult child of a parent with Alzheimer's. You need to keep in mind the clear definition of the customer as you develop your positioning. Knowing specifically who you are positioning the brand for is paramount.

Brand Opportunity Distilled

As we begin to work on determining our positioning of the brand, it is important to bring together and distill all of your work and analysis to this point. You need to synthesize your knowledge. This is best done by organizing your information into three major categories: competition (from chapter 1), customer insights (patient, doctor, payer, all stakeholders) (chapter 2), and brand (from chapter 3). This is essentially the first portion of your Brand Plan Rx Cohesion Map.

The positioning should be aspirational, built on deep customer insights/market understanding, and differentiated versus the competition. Distilled understanding and analysis of the market are critical inputs to positioning, concepts that we have covered in previous chapters.

The more intimate your understanding of your target, the more likely you will be able to identify a positioning that will motivate and drive relevance in the lives of patients and HCPs.

Building Blocks

A brand positioning is developed by using foundational building blocks—concepts that help us to synthesize our information and articulate our thinking. A common framework is the Premise, Promise, Proof construct. The Premise is the insight or the tension of the problem or unmet need/opportunity in the market. The Promise is the benefit, or how your brand will solve the problem articulated in the Premise. The Proof is the reason to believe (RTB): the evidence, the data, and the key information.

Using Nexium, a proton pump inhibitor that decreases the amount of stomach acid, as an example, let's deconstruct its positioning in terms of Premise, Promise, Proof.[15]

- **Premise:** Pain and discomfort from gastroesophageal reflux disease (GERD) may also be damaging the lining of your esophagus

- **Promise:** healing purple pill (you not only feel better, you are better)

- **Proof:** healing data

All the stakeholders in your marketing, sales, medical, and customer experience communities should have a clear understanding of the positioning. In this example, when you think of Nexium, you think of healing.

Although your positioning statement is just for internal use and not used promotionally, it's important that your brand's Premise, Promise, Proof outline resonates with your target patient. Later in the chapter we discuss different approaches for testing your positioning.

The positioning statement is for internal use and is a strategic tool that will inform the creative development, communications, tactics, support programs, branding elements, and so on. Many criteria are utilized to assess whether you have built a healthy positioning. We like to use the acronym BUMPS: that is, if you build a believable (B), unique (U), motivating positioning (MP) that sustains (S), you have created a healthy positioning for your brand. It is easy to remember because your brand "bumps" any current or future competitor out of the way with your positioning.

By considering the following criteria, you will ensure an impactful positioning for your brand:

- **Is it believable?** Based on the bedrock of proof, the RTB, the brand's evidence will answer the promise of the positioning. If consumers don't believe the promise, then you cannot create loyalty.

- **Is it unique?** Does it have a clear and distinctive place in the category? This is a place that no other product holds, as of yet, in the marketplace. Dove's "Every Woman Is Beautiful" positioning and campaign were unique to the category (i.e., toiletries, especially soap).

- **Is it motivating?** The positioning should harness a clear insight, one that activates the target customer. You do this by identifying a penetrating insight—uncovering the "deep *why*" that the brand can fulfill. This will unlock a compelling positioning.

- **Is it sustainable?** Have you assessed the future state of the market, competitors, new data, scientific innovation, and other factors? Have you looked at your own life cycle and indications to determine if the positioning holds true after the launch?

- **Is it single-minded? Have you been ruthless in focus?** Focus, focus, focus. You should be able to articulate the *one* core reason someone will want to use your brand.

- **How do you get to a compelling position?** First, identify the core insight: deep discovery of what the key stakeholder *needs* (chapter 2). Then review what the brand *has* (chapter 3).

Needs/Has Venn

One aspect of determining which insight is *the* insight that your brand should focus on is the marriage of *what a stakeholder needs* to *what a brand has*. As we've mentioned, part of the job of a marketer is to make decisions. Thinking through the stakeholders' needs and how they match what you can offer as a brand is critical. If you put the stakeholder first, at the heart of your decisions, you are then able to ask yourself how you might answer their needs in a way that another brand has not yet identified or can't deliver. This is the most difficult part of the decision-making process.

NEEDS/HAS VENN DIAGRAM

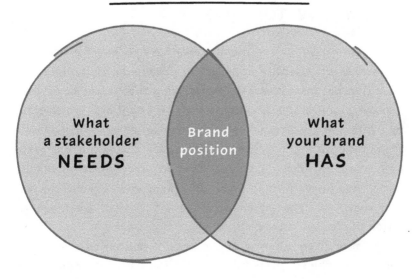

Let's Get Real

Many folks overdo how they determine what their brand should stand for in the eyes of their target. Lots of time and money is spent on positioning. We use some conventional tools, including benefit ladders; the Premise, Promise, Proof construct; and a positioning template. All really come down to the Venn diagram above. What does your consumer/customer need that your brand does? What is the role the brand plays in that person's life? By understanding the environment, the customer's needs, and the brand's superpower, this should be clear.

CASE STUDY: TOPAMAX

For Topamax, one of the most innovative solutions for migraine prevention, the scientific and medical stakeholders felt it was very clear that the insight was about reducing the frequency of likelihood of migraine attacks. Interestingly, while patients are plagued by the number of migraine attacks they suffer from, that wasn't an unmet need that was worth changing their behavior. They were used to dealing with migraines as they arrived; they had "personal hacks" and prescription products if they needed them. They felt they knew when the migraines were going to happen, so they preferred to stay the course with their current regimen.

The brand team looked at the patient journey of a migraine sufferer. All the sufferers talked about the days they would miss because of a migraine. Some of those days were really important, and people with migraines felt like the migraine hijacked their life. After completing a data dredge, the brand team realized that Topamax actually reduced the number of migraine days. The brand's "superpower" was greater control of migraine days, *not* preventing the number of attacks. By leading with the patient, we were able to unlock an emotional driver that the brand could answer. Could other migraine prevention treatments answer this? Yes, but no other brand had claimed this space and revealed such a compelling way to represent *what the brand has* to match *what a patient needs*.

Benefits Brand Ladder

Some folks are a fan of the "benefits brand ladder," a tool that allows you to methodically articulate and align your brand's value. We discussed features, advantages, and benefits (FAB) earlier; here we take that concept to the next level. We rely on that critical assessment of inside-out thinking to unlock the positioning for our brand. So we begin with the brand's features and attributes, then progress to its functional or rational benefits, and end with its emotional benefits. This is called the benefits brand ladder. It is a journey from the practical, logical, and factual, climbing its way up the ladder to less tangible and more conceptual aspects. Think of it as way to depict your brand on a ladder, with the rungs representing steps along the journey. The first, bottom steps on the ladder are the attributes, features, and functional benefits. The middle rungs are the emotional benefits, the results and outcomes of the previous rungs. The final levels of the ladder are the top rungs, which truly get you to the "real" meaning of your brand. While very difficult to articulate, these are certainly important and provide a unique competitive advantage. These are the transformational and societal benefits: How will it change me, my life, society, and so on?

Research has shown that brand decisions are made, and loyalty obtained, based more on the emotional benefits than the functional benefits. This is true even in the health and wellness field. While physicians rely heavily on scientific data, they are still human beings full of emotions. Those who have a positive emotional association with your brand tend to prescribe it more and be much more brand loyal. In other words, the higher you can move your customers up the brand ladder, the better your brand will perform.

Functional benefits are the attributes of the brand, the features, and the benefits. These are differentiated aspects supported by data and evidence. These attributes include such things as efficacy, safety, dosing, mechanism of action (MOA), and side effects. Usually they are found on the product label or in the package insert (PI). This first step on the brand ladder is of vital importance. You need to be on solid footing on the first rung of your ladder. Without establishing the brand's attributes, you will not have earned the right to move to the next rung.

Rational benefits are the consequences of the attributes to the patients, doctors, and payers. They build off the functional but articulate the *so what*. This is what the brand does. The rational benefits explain why the functional benefits matter to the customer. This could include such things as the result of better efficacy, fewer side effects, different MOA, and more convenient dosing, or could mean more adherence, better compliance, fewer withdrawal side effects, or shorter time to remission. This is where your brand begins to have meaning and differentiate itself in the eyes of your customers. Most pharma brands and marketers have become fairly good at this.

Emotional benefits are the result of the rational benefits; they build off them. This is what customers think and feel about the brand. These are the intangibles. The emotional benefits are not based on hard data or evidence but obtained from your insights research, from talking with customers, and from listening to what they are saying and not saying. Establishing its emotional benefits is where your brand can truly stand apart. Many pharma brands and marketers struggle with this level of customer engagement. Getting to this stage is critical if you are to realize your brand's true potential.

This gets you to your higher-order benefit: the meaningful emotional benefit. You need to continually ask the question *So what?* to get here. So what? So what? So what?

We prefer the needs/has Venn diagram to the benefit ladder, because it keeps it simple. However, we have outlined the benefit ladder to assist you in thinking through the ultimate benefit of your brand.

The essence of positioning is sacrifice. It is about making decisions. You cannot be all things to all people and still have a powerful positioning. You must be willing to give up something to establish a unique brand position. Yet, many marketers would rather cast the widest net, also known as "the everybody trap." They don't want to be tied down to a specific position because they believe it will limit their sales, or their opportunities. They are concerned about leaving opportunity on the table or niching their brand.

Positioning Template

When building your positioning template, be thoughtful of the intent, language, and simplicity of each box. The building blocks, benefit ladder, and needs/has Venn inputs give you the critical thinking tools to build a believable, unique, motivating, and sustainable positioning.

There are a number of positioning templates that companies use. We have used them all and found little difference between them. They strive to achieve the same purpose: to articulate in a succinct and clear statement the brand's positioning. We have found that it is more important to spend your limited time and resources on the content and gaining buy-in from your cross-functional team than on the format of the template. Sticking with the theme of the book, we use one that is simple, allows you to see the choices you are making, and shows the cohesion of your work to date.

The positioning template has four elements that, together in one statement, create the positioning statement.

Written out, the positioning statement would look something like this:

> *For* [target patient], *the brand provides* [single-minded proposition]
> *so that* [strongest emotional benefit], *because* [reasons to believe].

We begin with the *For* part of the sentence. This identifies the target patient. Who is this brand for? The answer should be based on your market segmentation and targeting work. The second part of the sentence is *the brand provides*: a single-minded proposition. It should be rational and emotional. This is the frame of reference, what the brand really is. Think of this as the competitive set. Next is the *so that*. This is the heart of the positioning statement, arguably the most important element. It articulates your brand's strongest emotional benefit, the single most important reason for the target patient to buy your brand. It should be only one benefit (not a list) that is based on your key insights and that solves a key frustration or tension for the customer. The last part of the positioning statement is the *because*. This addresses the RTBs, the proof, the evidence, the key attributes, the *why*.

A good test to apply to your positioning statement is the following two rules.

1 Your target in your *for* should value the primary benefit in your *so that*.

2 Your RTBs in your *because* should support your primary benefit in the *so that*.

Here are a couple of examples from the pharmaceutical industry:

Cialis

For couples suffering erectile dysfunction, *Cialis provides* a long-lasting effect allowing intimacy to happen spontaneously *so that* you can regain your relationship, *because* Cialis lasts for thirty-six hours.

Nexium

For GERD patients who are not getting better and are concerned about damage to their esophagus, *Nexium provides* fast relief and long-term healing *so that* you can feel better while protecting your stomach, *because* Nexium heals your esophagus lining.

Positioning your brand is a very exciting endeavor. As one of the most internally visible activities the brand team does, it often attracts much attention, even from senior management. As a result, you will get a lot of advice and divergent opinions and will find yourself trying to satisfy numerous constituents. With the best of intentions, mistakes are made.

We call these the deadly sins of positioning. We have seen these sins committed time and time again. Make sure that your team does not go down one of these paths. Use the following list of deadly sins as a tool to hold yourself accountable and to push back against those advocating for a positioning statement that falls into one of the sinful buckets.

(We have to quickly give a shout-out to a colleague, Erica Yahr. We wrote these deadly sins together and they share a wonderful place in many conversations of what to avoid.)

THE DEADLY SINS OF POSITIONING (OR, HOW TO MAKE YOUR BRAND'S POSITIONING GENERIC)

Best of Both Worlds
Maximizing the benefit, minimizing the risks
"We deliver strength... without compromise"

Kitchen Sink
Trying to communicate everything
"The powerful, safe, simple, convenient, and cost-effective solution"

Normalcy/Freedom
Overpromising the benefits
"We give you your life back"

The Gold Standard
Where to use the brand versus what it stands for
"First-line treatment"

The Go-To Brand
Your first choice/the only one you rely on
"The first and only approved brand"

Positioning Testing

Positioning testing is a subject that is very near and dear to our hearts. How you test something is as important as what you are testing. All the great thinking and research that went into building the positioning can be undone if the testing is not carefully thought through.

Here are five key things to get right when conducting market research that tests your positioning:

1 **Concepts:** Positioning testing (in its common form in pharma, Premise, Promise, Proof) is flawed. It is difficult to test stimuli that are not intended for external use. And we cringe in the back room when a patient, doctor, or payer starts wordsmithing versus looking at the intent, the concept of what our brand offers and if we are clearly representing the brand's best. We even provide a caveat to the market-research respondents: please don't dissect the language; look instead at the overall intent of what the brand has to offer. However, most people can't help themselves. There is a time and place to dissect the language—that's what message testing is for, when we translate the positioning into an external platform. For many market-research respondents, both consumers and HCPs, providing feedback on Premise, Promise, Proof statements is bewildering.

2 **Intuition:** Positioning should be informative versus directive. The best way to look at positioning is as a way to co-learn with your audience what is most motivating and relevant *to them.* Sitting side by side, with full transparency that our brand possesses a few superpowers, is the most authentic way of learning and gathering understanding to inform decision making. This is, of course, a qualitative endeavor, not a cold, quantitative, hard-number exercise. Having a quality conversation in which you listen and learn organically is the best way to build a positioning point of view. Sometimes this isn't easy in the health and wellness field. Health is intimate; it's very personal. Dignity, humanity, and empathy are

critical in a meaningful conversation with patients. For doctors, we use many different projective techniques as well as real-life conversations, journeys, and ethnographic stimuli to engage in a conceptual rather than evidence-based conversation.

3 **Comparisons:** It is critical to test building blocks that input positioning versus the positioning itself. What you should be testing is what the consumer needs and how the brand can answer those needs in a differentiated way (the needs/has Venn diagram). Include for contrast how the needs/has concept has been met by the competition, or simulate a future competitor, so it is absolutely clear that the overall benefit our brand provides is *unique*.

4 **Messages:** Begin with the end in mind. Translating the positioning to external copy is an unconventional way to validate positioning. Some teams truly feel they need quantitative or confirmatory direction before aligning on a positioning. If that is the case, one way to go is to translate the core of your positioning territories to key message platforms—actual copy that a consumer or doctor would see once the positioning is confirmed. This will require a bit more lead time as the agency will need to think through the specific language that embodies the positioning territories.

5 **Aspirations:** Your brand positioning needs to be many things. You are trying to meet a number of criteria (BUMPS) while still being concise and clear. In doing all of this, your brand's positioning needs to be aspirational. It should inspire your cross-functional team and agency partners. It should be a hope or ambition that is shared and that you are striving for: a destination that is just out of reach. It is a goal that is almost obtainable, something you are continually working toward. Most importantly, it's an aspiration that is built upon a deep insight from your target patient. And although you won't use the positioning statement externally or promotionally, it's a sentence that you know would result in an *aha* moment

for your target patient. They would read it and think, "These people really understand me." This methodology is highly recommended if you are shaping a market. It is very difficult to test a promise that is aspirational and has not established an unmet need.

Watch Out

We believe that for the purpose of positioning, your brand should be anchored on a patient insight, since the patient is the ultimate consumer of your product. You will, on occasion, encounter old-school thinking that sounds like this: "But it's the doctor who chooses which medicine to prescribe."

The prescriber is, and always will be, a key customer. But we know from our environmental pulse that their decision-making rights with regard to prescribing have been eroded over the years, based on hospital formularies and insurance limitations, among other things. So the statement above is not entirely true, and it is getting less true every year. Therefore, regardless of how specialty focused your marketplace is, it will be critical to form a positioning for your target patient, the consumer.

It is the patient who uses, ingests, injects, or applies your medications to get better and lead a healthier and longer life. This is a supreme act of trust that represents an enormous responsibility for every person who helps to get that medicine from development all the way to the pharmacy counter.

If you work for a healthcare company, we want you to stop reading this right now and go to your company's external website and look at the mission or purpose statement. Every single one of them refers to serving patients or helping people live longer, healthier lives. It does not mention serving physicians. If you doubt that patients are your ultimate customer, simply look at your company's website.

HCP insights are still critical. You will need to uncover them and use them when communicating your brand promise and your HCP message. Payer insights specific to your therapeutic area and class of medications are also essential, as we learned from the Praluent example in chapter 1. The team must constantly monitor that landscape for any new tidbits of information and feed them back to the cross-functional team as your strategy and messaging evolves.

If a benefit is distinctive but is not as important to our target stakeholder, we may need to sculpt the thinking of our prospects' minds. This is the definition of market conditioning, or disease awareness. We referred to this in chapter 3.

If you don't position your product, either your stakeholders or your competitors will position it for you.

Key Questions

- Have you followed the BUMPS framework?

- Is your positioning aspirational enough? Is it something that will always be just out of reach, never really obtainable?

- Does your positioning differentiate the brand in the marketplace? Will it hold a unique space in the target customer's mind and heart?

- Have you avoided the deadly sins of positioning?

- If you are a launched brand, can your cross-functional and affiliate teams clearly articulate your positioning? Do they buy into it?

- Do your current customer-tracker or brand equity results show gaps between your desired positioning and your customers' perceptions of your brand? What is causing that disconnect?

OBJECTIVES

Key Learning Points

4 Key Decisions

Source of Business

Objectives: The *What*

Cohesion Statement

BRAND PLAN Rx CHOICE MAP

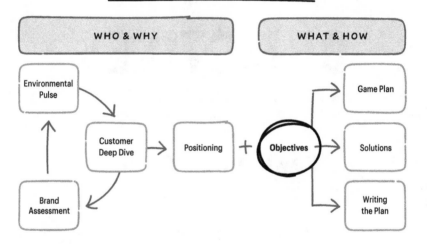

CONGRATULATIONS, you've just completed the first phase of the Brand Plan Rx Choice Map and are now entering the second phase. Think of the first phase as the analysis, the discovery, and the research part of developing a brand plan. The second phase is when we take all of the learning, understanding, and enlightenment and begin to make decisions—choices! Your mind and mentality should now shift gears, from curiosity and learning to focus and clarity. It is extremely important to build upon the analysis and work from the first phase and to ensure that you are applying those learnings. While the first phase was more iterative, this second phase is a more linear, step-by-step process. Each set of decisions will form the foundation for the next.

Your brand objectives are the first step in the second phase of the Brand Plan Rx Choice Map. They are a focal point in your brand plan, bringing together all your previous work and analysis into a clear and crisp proclamation of what you plan to achieve in the marketplace with your customers. To develop your objectives, you need to make some clear choices—decisions specific to your market, prescribers, patients, and payers. This chapter discusses these choices in more detail so that you can clearly identify and select the most advantageous and profitable opportunities for your business that match the customer needs. These key decisions that you make about your brand are the foundation for determining your source of business. With that information in hand you are ready to determine your brand and customer objectives. You then summarize all of that in a simple and clear cohesion statement.

4 Key Decisions

In his book *Breakthrough Marketing Plans*, Tim Calkins illustrates the idea of answering some key questions about your brand.[16] We have adapted that concept, built on in it, and made it relevant and specific to the health and wellness industry.

To make strategic decisions about your brand you need to answer a few key questions about your market, customer, and brand. These are questions that will determine how you plan to move your business forward: Where will your sales come from? Where will new customers come from? How will you grow going forward?

The four key decisions you need to make are organized into the following categories:

1   **Market/disease:** Are you creating a new category, planning to grow the category, or planning to grow share of market (SOM)? Are you a disruptor in the marketplace or are you going with the trend? Are you carving out a new and distinctive place? Are you stealing market share from competitors and if so, which ones?

    Grow the category or gain SOM?

2   **Prescribers (doctors, nurse practitioners, physician assistants):** Are you intending to increase the number of prescribers who write your brand or are you planning to have your current prescribers use more of your product by increasing the number of prescriptions per HCP?

    Get more doctors to use your brand or get the current doctors to use more of your brand?

3 **Patients:** Are you trying to stimulate awareness, get patients to seek treatment, encourage them to fill or refill their prescription, increase adherence to your brand, ensure better compliance, or create more repeat use?

Increase awareness, trial, purchase, usage, refill, adherence, compliance, or loyalty?

4 **Payers (health system):** Do you need to get your brand on more formularies (drug lists), do you need to get a better price approved, do you need to offer more rebates, or do you need better formulary status?

Improve access or increase your margin?

You need to have a clear answer for each of these four key questions. The answers will help you to determine your objectives, game plan, and solutions. Without a solid understanding of the answers it is near impossible to determine if a brand plan is good or bad, reasonable or far-fetched, sandbagging or overly optimistic. It is also difficult to develop and create your plan without answering these key questions. You cannot develop a solution unless you are clear on what you are trying to achieve in the marketplace with that solution.

Before you make these decisions, be sure to analyze the various scenarios and options. For each of these it will be important to make some calculations and quantify your analysis. Making some assumptions and doing the math is key. While you shouldn't be driven simply by the numbers, and will need to use some business judgment, the math will drive your decision-making process and help you to quantify your objectives and business results.

We will now go into each question in more detail so that you can determine how to "do the math" for each scenario and whether it's a strategic fit for your brand.

Decision 1: Market/Disease

You need to analyze and determine what provides you with the bigger business opportunity. Is it increasing the size of the category or is it increasing your SOM of the category? A rule of thumb is that if you are the market leader then it behooves you to grow awareness of the disease state; in turn, you will benefit disproportionately relative to your competitors. If you are the SOM leader, assuming the SOM stays constant, then growing the category usually ends up being a good thing for both patients and your business.

Not so fast. This is often the case, but not always. You need to do the math as well as some analysis to determine just how much time, money, and effort is required to grow the category. You then need to consider and weigh the same thinking regarding SOM. On the one hand, you might be the market leader with a commanding SOM lead but growing the category might not be the optimal decision. What if the category has come close to reaching its potential, like type 1 diabetes, or is in a difficult market to increase, like depression? Growing the category could be very challenging. On the other hand, you need to determine how difficult or easy it is to increase your SOM. How many competitors do you have? Are they relatively strong or weak? Are they gaining share or losing? Do they have any innovations like a new indication or delivery device? Do you anticipate new competitors?

There is no rule to apply here. You need to do the analysis and run the math in order to determine which decision has the greater opportunity. One approach is to determine what kind of investment, time, and energy is needed to grow the category by 1 percent and what that will net you, versus the investment, time, and energy needed to grow the SOM by 1 percent and what that will net you.

In addition to optimizing your resources, another reason to make the correct decision here is that each scenario requires different tactics. If you are growing the category, you are developing tactics designed to increase awareness, educate, and inform. This is a good decision if you have a novel mechanism of action (MOA), a new channel, or a new diagnostic. Typically, this is wide open space with little resistance. If you are trying to gain SOM, then your tactics are focused on building differentiation versus your competitors. It is a zero-sum game. You are competing and will face some blowback from your competitors.

One thing for certain is that the following equation is always true: the only way to grow your brand's sales is either by increasing the size of the category or by gaining SOM. The fastest way is to grow both. However, trying to do both is a common mistake many brands make. Every brand has limited resources and time. Ironically, wild success in one often results in a spillover effect in the other. A clear decision on one or the other will drive focus and result in initiatives and tactics that are fully aligned to your objective. Remember, writing a brand plan is all about making decisions and bringing cohesion.

Growing category or gaining SOM?

Category Sales × Product Market Share = Product Sales

Decision 2: Prescribers

Are you trying to get more HCPs to prescribe your brand or are you trying to get the current prescribers to prescribe more? In other words, are you better off getting a new customer (doctor) to begin using your brand on patients or are you better off getting a current customer (doctor) to use your brand more often, by prescribing it to more patients? More simply, ask how many doctors write your brand versus how often they write it. The only way to increase sales of your brand is to increase either the number of doctors writing scripts for it or the number of scripts each doctor writes.

This is also known as increasing the penetration rate or the buying rate. The penetration rate is the number of customers prescribing your brand. The buying rate is the average number of patients a doctor has over a period of time, and how many prescriptions a doctor writes per month.

In order to increase your penetration rate you need to find new doctors to use your drug. Concentration curves are a valuable set of data to have on hand to make this decision. The key information to know is the commercial potential for the next doctor on the list. How many appropriate patients does the doctor see in their practice? Knowing where you are along the concentration curve will help you to accurately determine the incremental benefit of the next prescribing customer.

Increasing the buying rate means getting your current prescribers to put more patients on your brand. This depends on the current usage and the remaining potential patients. This is fairly easy to analyze by looking at the current number of prescriptions per doctor and, based on the SOM with that doctor, how many more potential patients are in that practice. You will need to forecast how many appropriate patients your current prescribers can consider as potential patients.

Note that we are not implying anything unethical here. When we suggest acquiring new patients, getting doctors to write more prescriptions, finding new patients, and so on, we are not suggesting you partake in any improper behavior or tactics. All of your marketing strategies, efforts, and initiatives must strictly follow the good promotional practices set out by the US Food and Drug Administration. Absolutely everything that a health and wellness company does needs to be in complete alignment with FDA regulations and guidelines. All promotions and communications must be on-label. In the above example, for instance, we are not suggesting that you encourage HCPs to write prescriptions for inappropriate patients. While HCPs are able to use products off-label, *your marketing strategy and promotion should not.* Your brand plans need to be ethical, moral, and in complete compliance with regulations and guidelines at all times.

Watch Out

Trying to increase your penetration rate and buying rate simultaneously is the worst-case scenario. At the end of the day, you will achieve neither. Often the sales force doesn't pay much attention to your analysis and strategic decisions, as they are more interested in getting a sale and making a plan. They aren't too concerned if that is sourced from an existing or new customer. It is likely that the sales force will say "we can get both." The problem with this is that their efforts will not align with your marketing and medical efforts.

It is important to make these key decisions because your marketing solutions will differ considerably depending on what you are trying to accomplish. Finding new doctors to prescribe usually involves broad marketing, high-value incentives. Getting a doctor to use your brand involves solutions with a very targeted effort and incentives such as rewarding loyalty or size of purchase.

Increasing penetration or buying rate?

Penetration (# of customers) × Buying Rate (frequency of purchases) = Sales

Let's Get Real: Do the Math

A real-world example of a "do the math" scenario occurred shortly after the recent launch of a brand into a very large and competitive market. The rapidly growing disease state was already segmented with a number of classes of therapy to treat it. This brand was launching into the newest therapeutic class, which was small—only about 10 percent of the disease state—but was growing rapidly. The new category was dominated by one large brand that was well established, sufficiently resourced, and had an 80 percent SOM of the new and small class.

The newly launched brand was a tremendous success, rapidly stealing SOM from its one major competitor in the new category. First-year sales exceeded the plan and the trend lines looked promising; momentum was building. The brand team had aggressive and bold plans in place for the second year. The questions were (1) how realistic were the plans, (2) how would they convince the rest of the organization, especially the sales force, that the high expectations were achievable, and (3) where specifically would the sales come from?

The brand team had to do the math: Was it better to expand the new class or continue stealing SOM from the established brand? Growing the class meant finding new prescribers. This appeared promising as the class was new, had a small share (10 percent) of the disease state, and was growing rapidly—lots of tailwinds and promise. Stealing SOM within the class meant getting prescribers and patients to switch brands and going toe to toe with a major competitor.

Interestingly, when the brand team did the deep dive and analysis into the opportunity, the findings surprised everyone. They determined that the best way to grow and achieve their plan was to double their focus and efforts on their existing high prescribers rather than expanding their customer base. The math showed that depth versus breadth in their prescribers was the winning approach. The analysis showed that increasing the number of patients the current prescribers had on their brand by 20 percent would yield more sales than if they expanded their customer base by the same percentage, 20 percent. Increasing depth was an easier endeavor at this stage of the life cycle and required a much smaller investment than a breadth play. By doing the math, the team identified a key component of their winning game plan.

Decision 3: Patients

Determining where your patients are on the adoption curve is important. Refer back to your patient journey and key moments of truth (MOTs) here to get a sense of what you want to focus on. Every customer relationship starts with awareness and then moves to trial, usage, and hopefully adherence. Patients who seek multiple refills could be considered brand loyalists, although changes in insurance plans might force them to switch medication brands. Ideally, every patient moves along the curve and progresses from beginning to end. In reality, some are never aware of your brand, and some trial it in the form of samples from their doctor but never fill a prescription. Obviously, everyone moves along the curve at a different pace and at various stages.

For your brand you need to determine what single point is most important for your target patient. It is impossible to do everything at the same time. However, you can change over time and move along the curve at the same pace as most of your target patients. You need to remain focused and approach one phase of the adoption curve at a time.

Some brands, especially new ones, need to increase awareness. Other brands need to increase adherence. Perhaps patients using your brand fill the first prescription but for some reason fail to fill the second prescription.

Each stage of the curve requires different strategies and solutions. In truly general terms, TV commercials are great at building awareness but not so effective at building repeat usage. A coupon with your prescription is effective for repeat refills but terrible for driving awareness. The last thing you want to do is invest money in an after-purchase coupon that is designed for repeat purchases when what is needed is an initiative that increases awareness.

The real magic here is to understand the context and customer deeply through their disease journey. We need to apply the math/quantitative assessment against the market, epidemiology. Once we understand the prevalence, incidence, and opportunity audiences, we can build an attribution assessment and ecosystem plan which will weigh the addressability, lift, and potential conversion to make clear choices about the patient target metrics and goals.

Outside of the United States, direct-to-consumer (DTC) restrictions will limit your ability to impact the patient directly. However, there are still effective and compliant ways to reach your target patient. It's a mistake to assume that patient marketing is just the purview of the US market. We would challenge all teams to explore their options with unbranded messaging, patient advocacy, and product-specific patient information delivered through HCPs.

Increasing awareness, trial, purchase, usage, refill, adherence, compliance, or loyalty?

Decision 4: Payers

Striking the appropriate balance between access (potential volume) and margin (list price minus net price and cost of goods sold [COGS]) is a very important driver of your business. Variables to be considered when making these trade-offs include an understanding of your disease state, attitudes toward treatment, current treatment paradigms, competitive situations, innovation provided, current formulary status, and development and manufacturing costs. Ideally, you would make sure that your brand has "open" access to all patients and doctors who are in need and that they can obtain it at a fair market price.

The world of pharmaceutical access, pricing, and reimbursement is incredibly complex and constantly changing. In addition, the circumstances are different in the private insurer market than in the government system (Medicare, Medicaid, Veterans Health Administration, and Indian Health Service). The differences multiply when we look at the markets across the globe. For the purposes of the brand plan we will focus on the US market and look at access and margin from a strategic perspective. We'll leave the pricing negotiations to the account representatives.

Access to your brand for your patient and your doctor is the priority. Access at what price is the challenge. To take one extreme, you could have full access for all potential patients by giving the product away for free. Vaccines often operate with this aspiration in mind—governments and agencies are essentially paying, but they are free to the patient at the point of purchase. At the other end of the spectrum, you could price your innovative cure for a rare disease at such an astronomical price that no patients would be able to obtain access to it.

Finding the right balance and determining how much access you need relative to the potential market is when things become tricky. Access includes being on formulary for the type of insurance that various patients have. Sometimes access is restricted by prior authorizations, which means that the prescriber needs to go through some additional administrative steps in order to qualify the patient for the brand. Some brands have step edits, which means that patients need to fail on generics or less expensive brands first before they qualify for the brand. Some brands are "tiered," which means that the system prioritizes certain brands over others. The prioritized brands are not always the cheapest; they are often the ones that offer the largest rebate to the insurer and the pharmacy benefit manager (PBM). All of these decisions are negotiated and determined based on a number of factors, such as what the competition is doing in the marketplace, the therapeutic innovation, supply and demand, cost of manufacturing, and the unmet need in the marketplace.

You'll need to work closely with your payer colleagues to determine the percentage of the potential population that needs to have access to your brand relative to your competitors. For example, having a step edit might not be a deal breaker if your main competitor is subject to the same or more stringent constraints and if most patients are already on a less expensive, older brand.

Margins are vital to the sustainability of your brand and your company. Without achieving a positive and healthy margin from the sales of your brand, you won't be in existence for very long. How much margin depends on a number of variables and should be informed by your finance and payer teams. You can increase margin by either increasing the price or reducing the rebates offered, the manufacturing costs, and/or the operating expenses (OPEX) of sales and marketing. Your R&D costs are a sunk cost, as you have already spent the money to get the asset to this stage of development. For the purposes of this section of the brand plan, we will focus on two main levers: increasing the price and reducing the rebate. Both actions come with a trade-off relative to the amount of access.

For the brand plan, you need to determine if your priority is to increase the access to your brand. More specifically, is it to eliminate a prior authorization, reduce a step edit, and improve your tier? Once you have made the strategic access decisions you then need to work with your payer and finance colleagues to determine how much margin leverage you have to achieve your access goals. Keep in mind that the reverse can apply as well; you can determine that your access status is very competitive and that you would be willing to give up some access to obtain a greater margin.

A final note: The access situation depends heavily on the value proposition and the demand for your brand. The stronger they are the more likely your brand will obtain favorable access. This is especially true if you have health outcomes studies that validate your claims of improved patient outcomes or cost offset. Determining the number, timing, type, and size of health outcomes studies is a critical brand plan decision and could be a key solution in your brand plan.

Improving access or increasing margin?

Source of Business

The answers to these four questions determine your "source of business." Essentially, where will your business come from? From which point in the treatment algorithm? Stealing from which competitor? Targeting which segment? You need to consider this in the context of the current state, within the one-to-two-year brand plan time frame, but also with an eye to the future, in anticipation of market dynamics and changes. You need to make the optimal decisions for the short term while leaving future opportunities for growth open.

We call this our "go-to-market" strategy. It is the cascade of decisions that need to be made. It is part science and part art, so some elements will need to be quantified and some qualified. This will require referencing back to much of the work and analysis you have already completed. While each decision impacts a different aspect of your business, they all need to "hang together," complement one another, make logical sense, and link clearly to your brand opportunities. These decisions are evidence based but enhanced with a good amount of intuition and courage.

As we learned in chapter 2, the patient journey is foundational to understanding both the stated and unstated unmet needs in your therapeutic area. We discussed the need to take the many MOTs and prioritize them. Because many of our healthcare brands operate in a B2B2C market, we need to consider all customers' roles, influence, drivers and barriers in the care path, and, crucially, the brand choice. While the patient journey is our spine, the payer dynamics and HCP choice drivers will also be critical inputs when defining our objectives.

We use the framework below to clarify the questions that need to be answered and to break them down into understandable and comprehensive buckets.

DECISIONS	TARGETS	ACTIONS
Source of business	Market/category	Create, grow, steal
Volume	Providers, prescribers/ influencers/places	Breadth vs. depth, number of users vs. frequency
Adoption curve	Patients	Awareness, trial, purchase, usage, refill, adherence, compliance, or loyalty
Access	Payers, health system	Availability, costs, margin

This model will work in most cases. We have tried to simplify the decisions and actions to be taken, but in doing so it's impossible not to lose some nuance. You can, of course, drive awareness and trial with payers and prescribers, not just patients, as an example. These levers are not unique to a customer type and do not act independently of one another. However, we believe that in the health and wellness business it is most appropriate to line up the source of business decisions with the market, volume with the prescribers, adoption curve with the patients, and access with the payers. The chart above brings clarity to the types of decisions you will need to make.

When you think of marketing, you often think of creative campaigns, consumer market research, and sales meetings. But you also need to exercise your quantitative skills and leverage all that you've learned in forecasting and business analytics. If you take one thing away from this chapter it should be this: *do the math!* We can almost hear the sigh of relief coming from those of you who studied chemistry, finance, or engineering in undergrad and felt distinctly uncomfortable as you read about insights. Finally, a section you can sink your teeth into!

CASE STUDY: THE SGLT2 MARKET

Put your key answers into a simple one-page chart that lists the key decisions you have made, separated by customer group. This "source of business" slide should clearly state where you plan to play and how you plan to grow your business. In the oral diabetes market, a relatively new category called the SGLT2 inhibitor class (sodium-dependent glucose cotransporters) serves as a nice example: new class, highly competitive and crowded market, a market being segmented with new studies and claims, and lots of marketing spend.[17] This particular brand is the fifth SGLT2 to enter the type 2 oral diabetes market with a new efficacy claim relative to renal protection.

After running the math, the marketing team's decisions might look something like this:

DECISIONS	TARGETS	ACTIONS	5TH SGLT2 TO MARKET
Source of business	Market/category	Create, grow, steal	**Steal:** gain SOM
Volume	Prescribers	Breadth vs. depth, number of users vs. frequency	**Depth:** with current HCPs who see patients with renal impairment
Adoption curve	Patients	Awareness, trial, purchase, usage, refill, adherence, compliance, or loyalty	**Awareness:** of renal protective efficacy in patients with type 2 diabetes
Access	Payers	Availability, costs, margin	**Availability:** launch product with access in key accounts, prioritize competitive access vs. higher margin

You should be able to quantify your decisions to support your logic. For example, if you plan to increase the number of prescriptions/prescribers, then you should indicate your current prescription/prescriber rate, your competition's rate, the potential increase, and how much you intend to increase it. It could look something like this:

You know that, nationwide, endocrinologists write an average of three prescriptions per month of your SGLT2 brand; however, the concentration curve indicates that the top 50 percent of prescribers write seven prescriptions per month of your SGLT2 brand. Your main competitor is achieving twenty prescriptions per month from the same top 50 percent of prescribers. In this scenario, you've done the math and quantified your opportunity; you believe that your brand can grow from seven to twelve prescriptions per month from this group of psychiatrists.

This level of specificity is extremely helpful to align your organization and to have a metric that you can reach for and hold each other accountable to. This is especially true when it comes to collaborating with the sales force.

We worked on a brand where the analysis informed us that there was tremendous leakage, with patients not filling their second prescriptions. Clearly, measuring the number of second prescriptions was a lead indicator for success, so we made sure our marketing efforts were focused on just that. We measured second prescription fill rates on a weekly basis and by district across the nation. With the appropriate focus and adjustments to our messaging and marketing materials, we were able to increase the number of second prescriptions by 20 percent, which grew our SOM and achieved our business objectives.

Based on the knowledge and information obtained from your analysis in the first four stages of the Brand Plan Rx Choice Map—environmental pulse, customer deep dive, brand assessment, and positioning—you are now able to make the four key decisions. The answers will drive your brand's strategic questions.

CASE STUDY: THE CARDIOVASCULAR MARKET

The cardiovascular market, specifically the hypertension market, is a great example of how to apply all four of these decisions. Look at the market first. Are you trying to grow it, that is, increase the number of people seeking treatment for their hypertension and/or the number of people being treated for hypertension? Or are you trying to grow the hypertension market, meaning, get more people to seek and obtain treatment for hypertension? Or are you trying to grow a particular class of the hypertension market, such as calcium channel blockers, ACE inhibitors, or beta blockers? Or are you trying to grow your SOM within a particular class of hypertension treatment?

Second, take a look at the providers/prescribers. Are you trying to get more doctors to prescribe your brand? If so, which ones—general practitioners or cardiologists? What geographic locations, what size practices, hospitals or clinics? Or, are you trying to get your current prescribers to use more of your product? If so, is it to more potential patients or in place of what they are currently using?

Third, look at the patient. Are you wanting them to seek treatment, are you wanting them to fill their prescription, or are you needing the pharmacist to engage patients at the pharmacy, so they understand the administration and become more adherent? Do you need to communicate to patients about potential side effects so that they do not discontinue treatment? Do you need patient reminders so that they fill their second prescriptions? Do you need coupons and patient assistance programs to make it more affordable to the patient at the point of purchase?

The fourth and final decision entails taking a look at the payer space. Are you trying to increase access, availability, or volume? Or do you need to increase margin from your current volume? As in the previous examples, this formula is always true. If you increase access you will surely need to give on margin. The converse is also true; if you increase margin it will likely be at the cost of access/ volume. In the cardiovascular example, does your competitor in the same class have preferred status with the payer? Is another class cheaper and hence preferred to your class? Is your competitor beating you in the marketplace with rebates to distributors and prescribers or are they offering coupons to patients at the point of purchase? Do you need to increase your rebate or can you perhaps make up some margin by reducing the amount/percentage of rebate? Can you strike a deal that as the volume goes up, the percentage of rebate will increase?

Once you have determined the answers to these questions you are ready to begin developing your objectives, which will lead into your game plan (chapter 6/step six) and in turn your solutions (chapter 7/step seven). They should all be aligned to the answers you have developed.

Remember, we have intentionally simplified these concepts by assigning them to a specific customer type. You can, of course, increase awareness and trial with HCPs, or penetration with consumers (patients). And by increasing penetration, you are by default increasing brand awareness. As we mentioned, these decisions don't act in isolation of one another.

Marry the Math with the Mindset

We always start with the target segment. Thinking outside-in first defines the reality of our brand's potential.

1 Future market size and potential of each segment—are we driving demand, galvanizing influence, persisting with our target segments?

2 What are the current need states of those segments? Who are the stakeholders who will influence our core targets, positively or negatively? What will it take to drive their choice, adoption, loyalty?

3 Are there policy, health system, epidemiological, health burden, pricing, or care path considerations we need to address, mitigate, or motivate?

Then go to the inside-out aspects:

4 For an in-line brand, what is the current equity, are we continuing a positive momentum, are we mitigating any vulnerabilities? Do we have a new competitor to contend with or prepare for? Do we have new clinical studies, new indications to capitalize upon?

5 For a launch brand, think through the market you are entering and the superpower of the brand and what it will take to own the competitive positioning you have outlined.

6 What is the lift? How hard or easy will it be to achieve our competitive position in the market, and do we need to blunt current and future competition? Do we need to create a new mindset for the disease? And if so, how do we ensure credibility of that new concept?

7 Think through your internal needs. Do you have the capabilities, processes, partnerships, and human capital needed to drive success?

This is usually arduous, but we would recommend building tables and grids to critically assess these questions. Taking the time to do your homework will pay off when you are in front of your senior leadership team asking for the investment and defending your plan.

Watch Out

We have the luxury of writing this book without having to submit it to the same type of internal review your brand plan will undergo. We are talking about concepts and brands that we do not represent; therefore, we are not in a position to be concerned about future legal holds or depositions. However, your brand plan is always subject to those concerns. Our Watch Out note here is about language. We use the term "steal" to highlight a concept in our source of business chapter with regard to market share. Each company's legal and compliance teams will have a different viewpoint and tolerance level with regard to terminology.

Depending on recent legal judgments in the healthcare space, certain words that might have been acceptable in last year's brand plan will suddenly be banned from this year's plan. It is always a good practice to meet with your legal and compliance colleagues before crafting your plan to get guidance on good documentation practices and words that you can or cannot use in your plan.

Objectives: The *What*

As we did with the four key decisions, we have again adopted concepts and principles from Tim Calkins's *Breakthrough Marketing Plans* to be more specific and relevant to the health and wellness industry.

Now that you have made the four key decisions you are all set to write your brand's objectives. These are the goals that your brand efforts and actions are intended to attain or accomplish: the *what* of your brand plan. They are the desired end result. Many people get tripped up on the terminology. Is it a "goal" or an "objective," and what is the difference? We don't see much value in worrying about that, so we use the terms interchangeably. You should have two objectives for your brand: a brand objective and a customer objective. They should be clear, aspirational, and obtainable within the brand plan time frame.

1. Brand Objective

The first objective is the priority goal, which is pragmatic. It needs to be clear and financially oriented. This is inward facing—what you will achieve for your company or brand. It should be related to generating a profitable and sustainable business. Revenue is an okay objective, but profit is much better. The advantage of having revenue as a goal in a marketing plan is that the link to the marketing effort is quite clear. Revenue, however, is rarely an ideal goal. Revenue isn't what shareholders are seeking; profit is. The best brand plans would have an objective that clearly states both revenue and profit metrics.

In the pharma world it is very difficult for marketers to gain access to financial information regarding profit. That specific and detailed information is held very closely by people in finance. For various reasons it is not shared widely across pharma companies. This is a double-edged sword. On the one hand, if the people spending the money (OPEX) knew what parts of the business were more profitable, they would be better able to direct their efforts and activities. On the other hand, you are in the healthcare space and have a duty as well as a responsibility to help as many people as you can, and your decisions should never be solely based on financial incentives.

What you need for this part of the brand plan is one clear and concise financially oriented primary objective. If you are not able to obtain specific profitability information, you can still communicate a financially oriented objective in a number of ways. You can do this by articulating year-on-year growth or by using other variables as a proxy for profit. An example could be "Increase revenue by 10 percent and decrease sales and marketing expenses by 5 percent in the next two years."

2. Customer Objective

The second objective should be overarching, aspirational, and moti-vational. Usually, financial goals are not sufficient on their own. A marketing plan should have one or two other objectives in order to build a strong foundational brand that will endure. These are non-financial goals that should inspire everyone working on your brand across all functions. They tend to be more outward facing, things that you will do for the market, the therapeutic area, the disease, and so on. They are also fairly altruistic. This objective will allow you to build a strong brand, an enduring business, through the creation of a differentiated brand and a strengthening of a business capability. An example could be "Improve net promoter scores (NPS) by 10 percent with our target customers in the next two years."

SMART Objectives

If they are to be meaningful and actionable, objectives need to be SMART: specific (S), measurable (M), achievable (A), relevant (R), and time-specific (T). They should also not be confused with your brand's vision. Your brand's vision describes what you want to achieve in the long run; it provides a sense of purpose and big picture and sets the business's long-term direction. It's what your brand wants to be when it grows up. It is much more aspirational—a destination that you might never reach. Brand objectives are more concrete and located in the one-to-two-year time frame.

Here are some brand and customer objectives taken from real brand plans, and our evaluations of them:

"Achieve $1 billion in sales in ten years."

At first look this appears to be a great objective and meets the SMART criteria. However, upon further review the timing is too far out into the future. While it is great that a time frame is included, it is not realistic in the timeline of the brand plan, which is written with a one-to-two-year outlook. Ten years is too far in the future to judge if the brand plan will be able to achieve this objective. It would be perfect if the time frame was more realistic and within the time of the brand plan (i.e., not longer than three years).

"Increase profit by 10 percent."

This one is very good. It would be better if it included a timeline. For example, adding "by the end of the second year of the brand plan" would make this perfect.

"Beat Humira."

While very motivating and inspirational, this is not a good brand plan objective. It does not meet the SMART criteria. The first challenge is in defining what "beat" means—is this in sales, number of patients, number of pills, number of prescriptions? It also does not include a timeline. Finally, it is hard to determine whether it is realistic and achievable. Humira has multiple indications and customer types. Is the goal to beat Humira in each of its indications or to beat Humira across all of its indications cumulatively?

"Become the #1 autoimmune anti-inflammatory in two years post launch."

This objective is not a good one as it does not meet all the SMART criteria. Specifically, it is not clear what "#1" means. As in the previous example, is this measured in sales, units, patients, something else? Furthermore, it is not clear who the competition is. "Become #1" is not specific enough and thus is open to interpretation. At the end of two years it would be difficult to determine if the objective had been met.

"Increase market share of the diabetes market to 35 percent over the next twelve months."

This objective is perfect, assuming that 35 percent is realistic.

"Improve customer satisfaction scores with
psychiatrists by 20 percent in two years."

This objective is very clear and strong.

"Increase NPS by 10 percent among specialists by improving the
customer experience within two years of launch."

This objective is very good and meets all of the criteria.

Cohesion Statement

We've now answered our four questions and identified our source of business. We've done the math and generated our objectives. It's time to bring all of this together into one brief summary: a cohesion statement. This is an aspirational yet practical objective to drive growth. Your cohesion statement is an internal statement that brings together your decisions and helps you make sure they all hang together, are logical relative to one another, and have a "golden thread" that aligns them to what you want to achieve in the market and with customers.

Note that this cohesion statement is different from your positioning statement. The cohesion statement is simply a compilation of the four key decisions you are making, strung together in one simple and clear phrase. The positioning statement may look similar, but it captures an entirely different concept and serves a different purpose.

Here is an example of a cohesion statement:

Brand X will [create the market, grow the market, steal SOM]; *it will do this by* [increasing the number of prescribers/breadth, increasing how many prescriptions each prescribers writes/depth] *and by* [increasing awareness, trial, purchase, usage, refill, compliance, adherence, or loyalty] *with target patients, and by* [increasing access, increasing margin] *with the payers.*

Recall our earlier example of a new SGLT2 inhibitor and the choices the hypothetical team made with regard to their source of business and customers. Their cohesion statement might look like this:

Our brand will gain SGLT2 SOM; *it will do this by* focusing on current HCPs who treat type 2 diabetes (T2D) patients with renal impairment, *by* increasing awareness of new renal protective efficacy among consumers with T2D, *and by* ensuring broad access with no prior authorization with 80 percent of payers.

The cohesion statement is something that you will carry over and transfer to your cohesion map.

Key Questions

- Have you done the math for the various scenarios?

- Have you collected data to support your decisions?

- What assumptions have you made?

- Does your sales forecast align with the decisions you have made?

- Are your objectives SMART (specific, measurable, achievable, relevant, and time-specific)?

- Do you need additional health economics studies to give your brand a stronger payer value message?

6

GAME PLAN

Key Learning Points

Alignment with Patient Journey

Choice Criteria: ACT MOR

Cohesion Criteria: MECE

Verb Repository

BRAND PLAN Rx CHOICE MAP

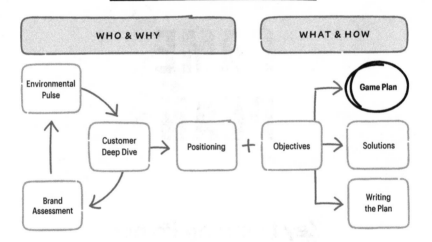

DEVELOPING A CLEAR and concrete game plan is the next vital step in the brand plan process. At this stage you have a great understanding of your market, your customer, and your brand. You have also established a positioning statement and a cohesion statement and have created brand and customer objectives. Now, how in the world are you going to achieve those objectives? That is where the game plan comes in. This is the *how*.

Many people call this the brand's strategy, or use more specific terms like "strategic imperative," "strategic intent," or "strategic initiative." Any of which is fine—use whatever term your organization prefers. Whatever you call it, we break the strategy down into three elements that together comprise the game plan. We like to call it a "game plan" because the term implies action, alignment, clarity, cohesion, and focus. Everyone can relate to the need for having a winning game plan. When teams win they usually credit two things: they say, "We had a great game plan" and "We simply executed better than our competition." When you have both you are unbeatable. This is true in life, sports, war, business, and even healthcare.

Your game plan is the heart of your brand plan. It is the *how* regarding the plan to achieve your brand's objectives. Your game plan should state in broad, sweeping terms what the brand will do to achieve the objectives. It sets the course, direction, and priorities. The game plan lands between objectives (what, outcome, results) and solutions (innovations, tactics, programs, activities). It bridges between what you want to do and the specific actions to do it. In other words, it is not the specific activity but the broad approach you will use to achieve the objective.

As we've discussed, an objective is *what* you want to achieve and a strategy is *how* you plan to achieve it. A sports analogy nicely brings this concept to life. In the football world, let's say that a team's objective is to win the game by two touchdowns and qualify for the playoffs. Their game plan might be to

1 leverage the offensive team's superior passing game to score touchdowns with the air game,

2 stop the opponent from scoring rushing touchdowns with their strong defensive line, and

3 obtain great field position with their special team's speedy kick returns.

In this scenario, some solutions under part one of the game plan (offensive passing game) would be the specific plays that the coach and quarterback call. That game plan makes a number of things clear: which plays should be called, which plays should *not* be called (for example, a running play), and the relevance and importance of making sure the offensive squad executes on the passing plays when the time comes. A clear game plan also allows for everyone to buy in and get aligned. If you agree with it, you are more likely to execute the solutions effectively.

It's not easy deciding what your game plan should be. First and foremost, it should allow you to achieve your brand's objectives. A framing we like to use is, take a step back and imagine this is your own personal company. You have done a deep dive into the potential of your brand and have outlined a clear objective in the marketplace. You realize that you need to invest wisely to achieve your objective. You also have a reality moment (gulp) that you will need to mortgage your house or take a personal loan to make the investment for your brand. What key actions do you need to take to ensure that personal risk is worth it?

That's the cold, hard choice mind frame you need to be in to ensure criticality of your actions. That reality check makes you think very carefully about the trade-offs you need to make and the gravity of your decisions when designing your game plan.

Alignment with Patient Journey

The best way to make sure that you develop an effective game plan is to look back at the great work you have done so far on your brand plan. The foundation of discovery, through your outside-in and inside-out thinking, has enabled you to clearly define your objectives and cohesion statement. Based on your outside-in and inside-out discovery, you identified how to achieve your objectives. Your game plan fuels these objectives. Remember, your actions serve as a link, a bridge, between your objectives and your solutions.

A great place to start is with your patient journey. Take a look back at this and especially the two or three key moments of truth (MOTs) you identified. These are a great clue in choosing the two or three strategies that should make up your game plan. With your MOTs you have already done the hard work of outlining the most critical and relevant aspects for patients as they navigate the disease. You have also ensured that the key MOTs are areas where your brand can have a positive impact.

Keep in mind you have done that work based on data, evidence, and market research, and with cross-functional input. If your strategies line up with the MOTs, you should feel very confident that you are addressing the most relevant and meaningful aspects as well as having alignment and buy-in from the broader team. Your game plan needs to align perfectly with your understanding of your marketplace, your customer, and your brand. Aligning to your MOTs will ensure this.

In addition, use the MOTs to validate the game plan. If one of your key MOTs is centered on the diagnosis of the disease—as it could be, for example, with diabetes—then perhaps your game plan should address that important time. Being diagnosed with diabetes is a very trying, emotional, confusing, and complex time for people, and one would assume that you would have a strategy addressing this stage of the patient journey and that your brand would bring relevant and meaningful solutions to it. Alternatively, if one of your key MOTs takes place when the doctor is making a decision on the treatment path and is selecting a prescription, then having a strategy that focuses on diagnosis will likely not make sense. Rather, you should have a strategy that is aligned to the treatment decision MOT.

To make things even simpler and clearer, if you have identified three key MOTs in your patient journey, then one would expect your game plan to consist of three strategies, each aligned to one of the three MOTs.

Choice Criteria: ACT MOR

When developing and writing your game plan, your strategies should meet some specific criteria. We use the acronym ACT MOR, because a well-devised game plan should get your team to act more. ACT MOR (explained below) will also help you to remember the criteria: your game plan should be action oriented, be clear and precise, have no more than three strategies, be measurable, be linked to your objectives, and be realistic.

1 **Action:** Your strategy should be action oriented. It will get things done, move matters forward. Each action should start with a verb, something to *do*.

2 **Clear:** A well-written strategy should be clear, precise, and specific, not vague, obtuse, or open to interpretation. It should state clearly what needs to be done.

3 **Three:** You need to be hyperfocused, with a maximum of three strategies. It might be tough to narrow it down to three or fewer, but choice is one of our guiding tenets.

4 **Measurable:** You should be able to measure all your actions, so that you will know if you are achieving them or not.

5 **Objectives:** Your actions should directly link to your objectives; they should fall out from the objectives.

6 **Realistic:** While they need to be both challenging and aspirational, your actions also need to be realistic. They need to be doable. Your team should believe that these actions are feasible and will allow your brand to achieve its objectives.

Cohesion Criteria: MECE

One of the biggest challenges in developing your game plan is in determining how general to be versus how specific to be. Obtaining the appropriate "altitude" is tricky. You don't want to be at such a high, general level that you are essentially covering everything and hence nothing. On the one hand, if you are too general, then any tactics would be appropriate and could align to one of your three strategies. On the other hand, you don't want to drill down and be so specific that you are limiting yourself to one tactic. For example, "Grow sales by launching innovative brands" is too general. This strategy is too vague and allows you to put a large number of tactics under it. Any innovation will work. Conversely, "Grow sales with patients who use co-pay cards" is too specific. In fact, it is so specific it is actually a tactic. Striking the right balance—flying at the right altitude—is tricky.

Using the MECE technique will ensure that the strategy in your game plan is just perfect—at the right altitude. MECE stands for mutually exclusive and collectively exhaustive.

Mutually Exclusive: Each item should be distinct, with minimal overlap.

Collectively Exhaustive: Together they encompass the entire range of possible outcomes.

"Mutually exclusive" means that when you stand back and look at all three strategies together, each one is distinct, with minimal overlap. "Collectively exhaustive" means that, combined, the initiatives should address the main issues and be sufficient to achieve the objectives. All possible strategies have been considered and covered. They should come from the opportune choices you have made based on your analysis of the brand, environment, and customer. You should be able to cover and address the entire landscape of your brand plan.

If your three strategies are MECE then you can be assured that you have developed a game plan with three distinct areas that are clear and do not overlap. They have very specific swim lanes. And you can be sure that your three strategies have appropriately covered the landscape and not left an obvious hole or gap.

Verb Repository

One of the most important criteria for developing your game plan is to have your strategies start with a verb. We have developed a list of verbs that work nicely in the health and wellness industry. They are listed below in our verb bank.

Use verbs that have meaning and are clear. Avoid hyperbole: grand words that are hollow and meaningless. They sound good but do not give people a clear understanding of what we want them to do. We call these "weasel words," and people use them all the time. Some of the typical marketing hyperbole we have seen includes words like "leverage," "maximize," "optimize," and "differentiate." They are safe and politically correct but way overused and not precise to your circumstance. Such weasel words would fit in any brand plan, but you can't afford any wasted words when developing your game plan—you need to be crystal clear. Nail the verb and set the tone for what you wish to achieve, which should line up with the boldness and tone of your objectives and the positioning of your brand. Verbs are the action driver. Below is a verb repository that provides example action words that are less business-jargony.

accelerate	challenge	gain	motivate	secure
act	claim	generate	negate	shape
advance	connect	hijack	own	stimulate
amplify	defy	incite	prevent	unite
block	demonstrate	inject	protect	upgrade
build	engineer	inspire	reinvigorate	
catalyze	excite	mitigate	revamp	

Some of these are quite aggressive. The words you end up putting to paper will, of course, need to be thought through from a compliance and intent perspective.

Listed below are some examples of phrases made from taking a verb from the repository and applying it to the health and wellness industry.

accelerate patient access	gain prescriber loyalty
advance the label	generate prescription rates
amplify the number of prescribers	inject sales force frequency
amplify the user experience with device	inspire prescription refill rates
	motivate new patient starts
block a new competitor	own a new indication
challenge new product development	protect the price
claim a caregiver program	reinvigorate the pharmacy program
	reinvigorate the category

revamp product adherence

revamp product costs

secure brand awareness

secure referrals

shape disease awareness

stimulate a product trial

unite advocacy

upgrade the digital mar-
keting footprint

We took a few of the verbs from the repository and developed some example strategies by applying the ACT MOR criteria:

- **Shape disease awareness:** Shape diabetes disease-state awareness by educating rural GPs.

- **Secure patient access clarity:** Ensure that psychiatrists and staff know how to navigate prior authorizations (PAs) for schizophrenic patients in need of an atypical antipsychotic.

- **Stimulate buying rate:** Accelerate the number of patients per prescriber who received filler for the top-tier dermatologists.

- **Build penetration rate:** Upgrade the number of oncologists who are comfortable using CAR T-cell therapy for non-Hodgkin's lymphoma.

- **Gain extended usage:** Inspire adherence by helping migraine patients fill their second prescriptions.

- **Generate new customers:** Double the number of Alzheimer's caregivers who seek evaluations for their loved one.

- **Inspire prescription refill rates:** Motivate people with diabetes to increase treatment adherence by completing their third refill for oral treatment.

A winning game plan is a key component of a brand plan. Your game plan should consist of three strategic elements that align to your MOTs. When writing these strategic statements make sure to apply the ACT MOR criteria. Together they should be MECE. The decisions made at this stage will have a dramatic impact on the success of the brand. It is a good time to pause, gain alignment, and make sure that you are 110 percent certain before moving on. Organizing a workshop or team meeting to gather input and confirm that your game plan is the absolute best it can be is a great idea.

Key Questions

- Does your game plan consist of three strategic elements that align to your MOTs?

- When writing the three strategies, did you make sure to apply the ACT MOR criteria?

- Together, are they MECE?

- If you implement your strategies, will you realize your brand's objectives?

- Are your strategies unique to your brand and circumstances, or can they also be appropriate to your competition?

- Is your cross-functional team fully aligned to these three strategies?

7

SOLUTIONS

Key Learning Points

Solutions: The *How*

Ideation

4 Ps of the Marketing Mix

Innovation

BRAND PLAN Rx CHOICE MAP

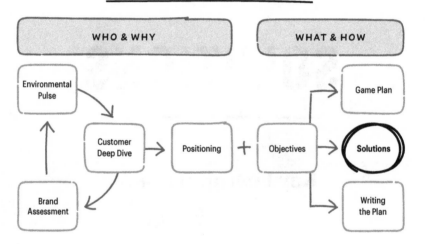

Solutions: The *How*

Let's start by clarifying what we mean by "solution." Our game plan is a synopsis of our actions, the most significant actions we need to take to achieve our objectives. The solutions are the way, the means, used to gain those objectives, to achieve those actions. Solutions are specific programs and/or initiatives that bring the game plan to life.

We've chosen the word "solution" purposefully. We want our brand to solve unmet needs of the patient, caregiver, and doctor—as they confront their disease or condition. There is inherent effectiveness tethered to the word "solution." We only want to identify solutions that will benefit the positioning we have identified for our brand. We only want to support solutions that meet the needs of our target customer at a key moment of truth (MOT). Solutions that do not do this should not be implemented or funded. We hope you can see, in doing this, the "golden thread" and alignment that this discipline will bring to your brand plan. There might be hundreds of wonderful, innovative, neat solutions. As we know, you can't do them all; you have limited resources and need to make trade-offs. At this point in the brand planning process, you need to ensure that you stay aligned to the "golden thread" you have created.

Too often we see excellent work in the analysis and brand planning process fall apart at the solution stage. A vendor comes in with a cutting-edge idea they have developed. A senior executive believes they have a brilliant idea that will save the brand and leave a mark on the industry. A consultant is motivated by something they have seen work for another client in an entirely different disease state. A young and eager MBA wants to prove their worth and is pitching a creative idea they have been incubating for some time. Or, our favorite, the medical team has come up with an idea based on their time in the clinic and from the advice of key opinion leaders (KOLs).

It is critical at this stage that you ensure your solutions are 100 percent aligned to the brand plan. Working it backwards is a great litmus test. Are your solutions aligned and do they fit into one of the three strategies that you developed in your game plan? If executed properly, will they achieve your brand's objectives? Do they act on a key MOT identified in your patient journey? Are they designed with your target customer in mind? Do they resolve a need, frustration, or concern that you identified in your insights work?

Solutions provide the next level of detail on the *how* and then build on it with a specific idea, activity, or initiative. They lay out precisely *when* and *where* things will be done.

You can ask some general questions to start formulating your thoughts. Why are we doing this? How will this make a difference for our target segment? How does it support our game plan, our objective? Who will implement this tactic? How long will it take to make an impact? How much will it cost? How will we measure it?

The formula below represents a logical way to think critically about your solution. Each solution should be explained on a one-page summary. This summary will explain the solution simply (so a seven-year-old can understand it), including information such as what it is, what problem you are solving, how it works, who does it, when it will be done, whom it will be done for, how often it will be done, what it will cost, the key customer insight you are addressing, the key customer indicator, and how it will achieve the objective. This forces focus and clarity. It is also an easy way to check for alignment and cohesion of your plan.

Here is a simple formula for a solution:

- **Title of the solution:** Like the title of a book, what is the title of the solution?

- **Brief description:** Describe the idea as if to a seven-year-old.

- **What problem are you solving?** What will we tackle, solve?

- **Key customer insight:** Why will this be effective for our customer? What moment in the journey is this solution intersecting with?

- **Key desired effect:** What will be the desired effect for the customer?

Great solutions have analytical rigor. This means you have thought through your budget, timing, and amount of activity, among other things. At this stage it does not need to be down to the penny, but you should have some idea of what things will cost, as well as how often you will do them, and so forth. Is it something that you will do twice a year like a big event? Or is it something that happens with every purchase, or occurs daily or weekly? Think through the frequency, scope, volume, and, eventually, cost. You should also tie the big idea back to your objectives. If you are spending a certain number of dollars it should be clear what results this spending will obtain and by when. This should be aligned to your previously stated objectives.

Ideation

Getting to innovative solutions. The most constructive way to build innovative solutions is to start with *Yes, and . . .* versus *No, but . . .* thinking. "Yes, and . . ." are the words you would use in a meeting. It is how you bring to life the idea of, first, divergent thinking and then, later, simplifying and choosing the best solutions, or convergent thinking. To identify the most critical actions, we employ divergent and convergent thinking models. Use divergent thinking first, by listing all the possible actions you believe could be necessary, before you decide on the most important solutions to achieve your game plan.

In this process, pull together a diverse team to brainstorm possible strategic solutions—path ideas that could propel and anchor your game plan. Say "Yes, and . . ." to all ideas. Once you have come up with a number of creative ideas, you then begin to choose, consolidate, eliminate, expand, and reconcile. The divergent part is creative, free flowing, spontaneous, and out of the box, with no wrong answers or ideas. This is when you utilize the *Yes, and* . . . phrase, not the *No, but* . . . one. Build on your colleagues' ideas. Instead of focusing on why they won't work, think of things you can include, do, or change to make them work.

The convergent part is the opposite. You need to get realistic and practical, narrowing down to a few good ideas. To progress from divergent to convergent thinking, employ the criteria you believe are necessary to prioritize your strategic actions. We like to brainstorm on all the possibilities by referring back to our journey map, looking keenly at the leverage points, the target mindset, our competitive position, and the headwinds and tailwinds of the environment.

Solution Day

Clear your calendar and create a "Solution Day." Bring your cross-functional team together for an ideation session. There are a few simple steps that will ensure this process is effective and leads to big ideas and solutions that will meet your criteria: that is, that are innovative, solve a problem, achieve your objectives, and fit within your game plan.

Preparation: Importantly, you need to digest and internalize the data. Ensure you have a deep understanding of your target audience, harken back to your journey map, and consider the environment, category dynamics, headwinds and tailwinds, payer choice drivers, and brand advantages. Create a snapshot overview of the most critical aspects to brief the team members who are coming to the category with fresh eyes. We would suggest assigning a dynamic presenter to establish the backdrop of your opportunity, as a brand. Start off with reviewing, whiteboarding, and outlining all of the information in an organized and methodical manner. Providing pre-reads is helpful for attendees who might not have been along for every step of the journey or who might not have been present for any or much of the market research. It would be great to be able to show some of the videos and recordings of the market research that has led you to this point. Consider inviting special guests—maybe someone from a local advocacy group, a prolific blogger, or a local professor who has done some interesting research—who align with your brand's disease state, your patients' needs, or your doctors' concerns. Your first goal is to get everyone up to speed, on the same page, equipped with all of the information.

See the future: Also, think about what the world will look like twelve to eighteen months from now. How will the market/category look, what dynamics are in play, are there new considerations to be prepared for? How has COVID-19 impacted your disease state and the healthcare system? What are direct consequences and unintended consequences of COVID-19? Some scenario planning and brainstorming discussions open the mind up to various possibilities.

The divergent thinking lens: Then bring in data, analogs, cultural codes, insights, or other category stimulus that will get your team to think differently about *how* to action the strategy. When you think about your objective, what kind of lens do you wish to bring to your team? Are you building a new category where you will need to form new habits and establish new values for choice? Are you disrupting an entrenched competitor? Are you gaining loyalty with physicians whom you haven't been able to win over yet? Are you driving demand more broadly? Think about what your objectives and strategic imperatives are screaming. This will unlock the kind of inspiration session you will need to unearth ideas—*big ideas*—to achieve your objectives.

Face any challenges with vigor: If you are mitigating a huge hurdle, face that hurdle and think through the *why*s of what your brand is facing. The homework and diligence on understanding how you can influence choice will open up clear and innovative solutions.

Sometimes it is helpful to bring in an inspirational speaker to spark new and fresh thinking: for example, a monk or a hostage negotiator. Sometimes it helps to create an experience. Have your team members try to order the competitors' products, or try to make an appointment with a specialist, or search online for some important information.

At one such meeting working on a product for diabetes, our lunch break was at a restaurant across town. We broke the team into smaller groups, each representing our target patient, and each was given an assignment to accomplish prior to arriving for lunch. One team needed to buy test strips at a pharmacy, another had a hypoglycemic event and had to stop along the way for orange juice, another had to find a place and time to test their blood sugar prior to lunch, and yet another was given a cane and special shoes for diabetic neuropathy and told to take the bus across town with their insulin in a cold pack.

In sitting on the board of Easterseals Crossroads, a nonprofit that assists people with disabilities, we obtained insight into a unique approach. Easterseals has a noble mission, limited resources, innovative minds, and fulfillment in what it accomplishes. We have learned of and seen firsthand some of the most creative solutions to human problems imaginable. We took one of our brand teams for a visit to Easterseals. After a brief tour we all volunteered in various parts of the operation: we helped clients, learned from employees, made a difference, and got inspired. In the afternoon we had an ideation session working on brand plan solutions and came up with ideas, thoughts, and concepts that we would never have realized on a conference call or a Zoom meeting or even at the corporate office in an all-day brainstorming meeting.

What's the current state? If you are writing a plan for a brand on the market, take stock of what you have accomplished, what worked, what didn't, and why. What would you continue, what wouldn't you do again, and why? What did you learn from employing that initiative? We suggest developing a Start, Stop, and Continue slide. Take inventory of your solutions from last year's brand plan and list the initiatives you should start, those you should stop, and those you should continue. The "continue" list could be adjustments, tweaks, and improvements based on your learnings and feedback.

This thinking served us well with L'Oréal, the world's largest cosmetics company. L'Oréal has been around forever. Women are faced with so much choice, confusion, and anxiety in buying cosmetics that they often just stick with what they know. L'Oréal's base of consumers needed to expand.

Strategy: In this environment, how do you get people to try new products?

Objective: Increase product trial with prospective consumers.

Game Plan:

1 Inspire new users to experience the "magical-ness" of L'Oréal makeup.

2 Personalize L'Oréal products to L'Oréal users.

Solution: An entirely new way to try on new makeup—the Makeup Genius app. This app, which allows users to see what various makeup products would look like on them, is a fun and realistic way to explore the beauty that L'Oréal provides. (It's the world's most popular beauty app.)

4 Ps of the Marketing Mix

Many marketing programs reference the 4 Ps—product, price, promotion, and place (distribution)—when brainstorming solutions. The 4 Ps are a framework for making marketing decisions, made famous by Philip Kotler in his marketing textbooks back in the 1960s. They are still as relevant today as they were back then. However, similar to how we felt about the SWOT analysis, the 4 Ps has its limitations and we believe that some other methods get to various solutions more naturally. If your company prefers to assess and organize thinking utilizing the 4 Ps format, then use the 4 Ps methodology. You could also utilize the 4 Ps as a pressure testing mechanism. Even better, once you brainstorm on your solutions, check those ideas against the 4 Ps, pressure testing any potential areas that need further strengthening.

Product can include a number of solutions covering areas such as quality, design, features, packaging, and various related services. In the health and wellness industry this can include new indications and line extensions (NILEX), additional data, publications, new formulations, and improved delivery devices. An increasingly important area of differentiation is in related product services. This includes such things as patient support programs, patient advocacy, digital solutions, and community and communication services. Also included under product solutions are a number of branding elements, such as brand name, logo, colors, and images.

It is important to refer back to chapters 2 and 3 when evaluating solutions related to your product. You need to make decisions about your product relative to the competition's. Some things might be table stakes (a ticket for entry); others might truly differentiate you, though some might be of no incremental value.

Price is an ever-increasing top priority in healthcare marketing. Ensuring access to and reimbursement for your brand could be a key tactic on which your brand will need to focus. In fact, price has become such a crucial element that pharma companies have created dedicated teams to this one P of the marketing mix. Many marketers have become pricing and access specialists, developing deep knowledge and expertise in managing pricing. This is arguably the most vital element of the marketing mix in today's pharmaceutical industry. We could write an entire chapter on pricing alone. For the purposes of writing a brand plan, we will only touch upon a few key aspects of pricing. The truth is, pricing decisions are usually raised to senior management, who end up making the decisions, leaving the marketing teams to manage, communicate, and implement.

Price is much more than the dollar amount that your brand is sold for. It begins with developing and defending the value proposition of your brand as well as obtaining the best and most appropriate price. You will need to determine both your brand's list price and its net price, as well as your pricing strategy for the commercial (private insurance) market. This will have an impact on the pricing strategy for the government market (Medicare, Medicaid, Veterans Health Administration, and Indian Health Service). You will also need to determine your global pricing strategy: the price you will sell your brand for in other countries. Each of these key decisions has implications for the others. For example, your commercial pricing strategy impacts your government strategy, as the government bases prices and discounts on the commercial prices. Your prices around the globe have implications in that many countries reference price, so the order that you launch and the prices that you negotiate can have a positive or negative effect on your overall business.

Without having tactics in place to ensure that your brand is getting the optimal price in the market, you will have a difficult time reaching your objectives. Additional tactics related to price are coupons, rebates, discounts, and numerous patient support programs. These are very important decisions that could make a huge impact on the ability of your firm to achieve your brand's objectives.

Promotion of your brand is a popular and very visible part of your brand plan. Tactics for promotion could include detail aid materials (sales brochures) for sales representatives to use with doctors during sales calls; advertisements in magazines, journals, and waiting rooms; digital and social media campaigns; and direct-to-consumer (DTC) television advertising. Developing a clear, concise, consistent, and meaningful message is key to the promotional strategy. All your communications need to align and reinforce the core messages in a consistent manner.

This element of the marketing mix is currently going through a tremendous amount of disruption, in the positive sense. The advent of smartphones, wearables, and social media is transforming communication to patients and doctors. As the concepts of "care anywhere" and "health literacy" are realized, the promotional strategy is following suit. Patients are expecting to receive care anywhere, meaning in various locations and settings and at times that are convenient and appropriate to *them*. Telehealth is just one feature that has increased tremendously as a result of COVID-19. "Health literacy" refers to the concept that patients learn, know, and understand more about their health than ever before. With the advent of technologic and digital solutions, many patients are getting their health data before their HCPs. Sources of data range from Dr. Google to health insurers monitoring your steps and calories in exchange for discounts to you seeing your lab results before you visit your doctor.

Place includes how you distribute the product and get it to the right place in the market at the right time. This could include distributors, wholesalers, pharmacies, hospitals, and clinics. For some brands it is fairly straightforward; for others, such as biologics, oncolytics, and other large molecules, it can be more complex and a critical success factor. Many products require special shipping and storage conditions. For example, many vaccines, including some for COVID-19, have very specific cold-storage and freezing requirements. Manufacturing falls into this category as well. Ensuring safe and reliable supply is a must.

Place is also where care is given. As mentioned above, as a result of COVID-19 this aspect of healthcare has gone through a rapid and accelerated transformation. New ways of making healthcare more convenient, safe, simple, and fast are emerging. The old days of a large, crowded hospital with long wait times are being challenged by services such as retail clinics, home care, and telehealth, which are all experiencing explosive growth.

Key Desired Effect

Measurement is a key element of your solutions. What will drive the desired behavioral shift in your target audience? When you look at these questions through the lens of customer-centricity, you can make sure your metrics line up with your intent and with the impact of your brands on the lives of your key targets. Think about new ways to measure the effectiveness of your marketing—it's not enough to know how many impressions you made on Facebook. The value of your relationship with your customer should be your benchmark for success.

All of your solutions should have a clear and simple metric: your key value indicator. The benefit of this metric is that the team, management, and you can determine if a solution is successful in your endeavors to achieve your brand plan. You should also obtain constant qualitative and quantitative feedback against your metrics. That way you can also determine what changes and improvements, if any, you need to make to your solution.

Key value indicators can become a powerful management tool. With the proper metrics in place you can determine whether you should increase the *frequency* of an initiative, increase the *reach*, or make other *modifications* to your tactic. If your feedback and measurements do not lead to obvious improvements in frequency, reach, or modifications that will improve the impact and return on investment (ROI) of your solution, it is time to save your money and cut the program. Either give that money back to the corporation or reallocate it to a program with a better ROI.

Here is an example. Let's suppose you are running a program that shares key clinical information with your top 20 percent of prescribers—some sort of educational symposia. You need to determine the ROI of that tactic and whether you should increase or decrease spending or reallocate those resources to other higher ROI priorities. Or perhaps you should make some adjustments and improvements that will make the program more attractive to the appropriate targets, have a more meaningful impact, and so forth. It is fine to make these adjustments (and there always are some), assuming that any tweaks to your program are minimal and you are able to make those adjustments easily.

Next, you need to determine if you should expand the program and invest more in this initiative. If so, do you increase the frequency—run more programs with the same customers, say, up from six per year to ten? Or do you increase the reach—expand to include more doctors, say, from the top 20 percent to the top 30 percent? The decision depends on a number of variables. Most important are the concentration curve and the amount of upside coming from your next potential customer. Does the next person you would invite have more potential than the first person you invited who is attending the event again? Is there anything new to present to the already attended customer? Would it be better to repeat the initiative if it is a complex topic?

Keep in mind that the opposite works as well. Let's say you have finished presenting your brand plan and you have general agreement on your strategy and tactics. In fact, senior leadership likes the plan a lot and has given you terrific support and approval. However, budgets are tight and the executive committee has asked you to reduce your operating expenses by 15 percent. How do you best go about that?

Do you just cut all your programs by 15 percent—the proverbial "haircut"? That would be a fairly crude approach and certainly not optimal. An informed approach would look at your metrics and feedback and identify ways to reduce either the frequency or the reach with the least negative impact on your performance.

Innovation

The healthcare industry is ripe for innovation and disruption. We have been saying this for a long time. More than a decade ago, bestselling author Clayton Christensen and colleagues published *The Innovator's Prescription*.[18] In it they famously articulate how healthcare and other industries will adopt, evolve, and innovate. On the one hand, this happens at a tremendous clip in healthcare on a daily basis. The use of technology, science, and big data is incredible and has resulted in amazing advances, from sequencing the genome to treatments that cure diseases once thought incurable to rapid advances in treating cancer. We have seen remarkable health advances around the globe with the United Nations Millennium Development Goals, which continue with the Sustainable Development Goals initiative.[19] COVID-19 has stimulated a number of advances, such as vaccine development in record time, an explosion in telehealth, new treatment paradigms, and unprecedented collaboration and sharing of information between public and private entities as well as between competitors. On the other hand, we are still stuck in the old paradigm where the sick patient needs to go to the doctor's office and sit around with other sick people for sixty-two minutes on average to get to see their doctor for an average of seven minutes; 43 percent of that time the doctor is busy entering data into their electronic medical records, spending only 28 percent of those seven minutes actually talking to the patient.[20] Much innovation and transformation is needed in the health and wellness space.

Modern Marketing

The healthcare consumer's expectations are changing rapidly and the healthcare industry needs to keep up. A few reasons for the changes in the healthcare customer experience (HCx) are the following:

- The Amazon effect: our expectations of seamless, quick, simple, and anticipated service has been changed forever

- Health literacy: Dr. Google has put information about our health at our fingertips

- Health data: we have access to our data at the same time as our physician—in fact, we often know our results before our doctor does

- Wearables: these allow for health data to be collected on a continuous basis

- Digital natives: as more and more patients and doctors who are digital natives become the providers and consumers, the use of technology increases exponentially

- The concept of "care anywhere" has taken hold

All of these trends take us to the "blue ocean" made famous by W. Chan Kim and Renée Mauborgne in their book *Blue Ocean Strategy*.[21] The days of the medical sales representative arms race are over. Brands are now coming to the market with solutions for patients and prescribers that are new, fresh, and innovative. The old "red ocean" of medical sales representatives dropping samples, TV commercials with happy smiling people on the beach, and rebates negotiated with pharmacy benefit managers (PBMs) will *not* create the brand success that you and your leadership are looking for. It might be that you need some of the old tried and true basics. However, they are not going to allow your brand to break through all of the clutter and become a success. You need to find a "blue ocean" and provide solutions that have not been tried in healthcare before. They could be truly innovative, they could be borrowed from other industries, or they could be customized for your brand and disease state.

Many of these involve technology, big data, and digital solutions. The concept of "care anywhere" has come of age. Telehealth is exploding. Patients won't accept the old brick-and-mortar healthcare solutions. Healthcare is being turned upside down as you are reading this book. As a brand manager you are either part of the revolution or on the sidelines watching it pass you and your brand by. Your brand plan needs to incorporate and anticipate these changes.

You need to ask yourself important questions that may uncover new ways to drive demand. How can you drive and influence decision making? How can you create value? Ideate around patient services and telehealth. Companies such as Livongo Health, Omada Health, and Cecelia Health have made tremendous inroads in developing new business models, innovative payment schemes, and novel services. How have you developed and adapted your promotional strategy to speak with and communicate to the telehealth doctors who are writing your prescriptions (and hope you are not sending a medical sales representative to the telehealth call center to meet with them face-to-face)? Many more ideas like this are brought to life by cardiologist Eric Topol in his book *The Patient Will See You Know*. Brand plans today need to create an Amazon-like HCx, by understanding that patients are more health literate than ever before, have access to data in real time, and are likely digital natives. Above all, you need to understand that many doctors are not comfortable with all of this change. Recall from chapter 1 that about half of the doctors in the United States are over fifty-five years old. How you approach this is critical.

Extensive Example of an Innovation That Would Impact Your Solutions

Partnering with disciplines, companies, and technical innovations is an inventive way to drive differentiation and outcomes. Remote health (telehealth, remote diagnostics, and video consults) and digital health companies are great examples of changes in the healthcare environment that are having a profound impact across numerous disease states and need to be incorporated into your brand's solutions. All these trends come together in Teladoc Health, the telehealth market leader that has been experiencing exponential growth especially since COVID-19 hit.

The company has established partnerships across various sectors of the healthcare ecosystem, from software to providers to insurance companies. For example, Teladoc acquired InTouch Health, the market leader in provider-to-provider virtual care (think of it as LinkedIn for doctors, with additional features specific to healthcare) as well as Livongo Health, a provider of virtual chronic care management programs. This vertical integration allows doctors to connect with one another on a secure platform and obtain advice, expertise, second opinions, and so forth through InTouch Health. It also allows them to complement treatment with proven well-being solutions and support, which are all virtual with the addition of Livongo. This nicely complements Teladoc's leadership position in consumer virtual care.

This is just one example of a rapidly changing landscape in a subsector of the industry that can impact the journey of disease, treatment, the doctor–patient relationship, and more. With COVID-19 emerging on the scene, Teladoc was in a perfect position to help millions of Americans get the care they needed when visiting a doctor's office was not allowed.

Digital health is another change that brand plans need to incorporate. Omada Health is a good example of a digital care provider that offers support for a number of disease states. It is challenging existing paradigms and delivering innovative solutions. One instance is its payment model. Consumers do not pay unless, and when, their health outcomes objectives have been achieved—a true pay-for-performance model. Another such company is WELL Health, which provides

AI-powered health engagement in the form of a concierge service to reduce costs and improve outcomes. WELL has partnered with employers and insurers to deliver customized health services, identifying patients who are in the most need and are driving up costs. It offers personalized care that improves outcomes while driving down costs.

These innovative solutions can have a profound impact on how you market your brand, transforming such foundational inputs as the journey, the choice drivers, the barriers, and the intersection and status of many stakeholders' relationships, as well as the way in which we marketers connect and influence stakeholders. For example, what is your strategy to target and communicate to telehealth doctors? Is it the same as or different from your current sales and communication strategies? How about using AI and partnering with any of the new telehealth and digital companies to ensure that when patients and doctors are communicating with the new medium, your advertisements are targeted and delivered? If a person with diabetes is having a televisit, you need to make sure that your diabetes brand has an advertisement that shows up on the screens of both the patient and the doctor. Alternatively, if a parent is dialing in regarding an ear infection for their toddler, make sure that your diabetes brand is not being advertised but your antibiotic is.

Key Questions

- Have you identified three to five key solutions that will truly drive the success of your brand?

- For each of those key solutions, have you developed a one-page summary that captures key information?

- Does each of your solutions have metrics, specifically, a key value indicator?

- Does each of your solutions clearly fit with one of the three strategies in your game plan?

- Have you anticipated the future when considering your solutions?

- If you had to pick one key solution that would have the most impact on the success of the brand, which one would that be and why? What if you doubled your efforts on that solution?

- If you had a 20 percent budget cut, what solutions would you eliminate or reduce?

WRITING
THE PLAN

Key Learning Points

Brand Plan Rx Cohesion Map

Assumptions, Risks, and Contingencies

Timeline and Milestones

Financials and Metrics

Storytelling

BRAND PLAN Rx CHOICE MAP

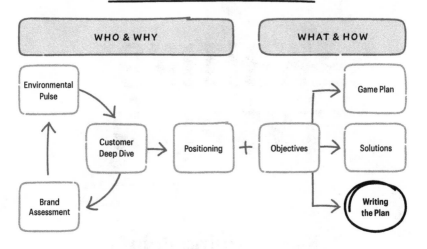

T HERE REMAIN a few loose ends that need to be tied up before your brand plan is complete. These final steps result in a cohesive, well-thought-through, well-written plan. The goal is agreement and buy-in to the objectives, game plan, and solutions. Now we need to know how feasible the plan is, how we can make it work, and how we can best tell the story.

Who Writes the Plan?

The brand plan should be led by the marketing team. Usually a senior brand manager is the lead person ultimately accountable for delivering the finished plan. However, this is a cross-functional team exercise. Alignment and enrollment of critical stakeholders along the way is incredibly important in building the plan. The core team should include sales, medical, market research, access and pricing, regulatory, manufacturing, public relations, and legal. The extended team should include general management and senior leaders. At the end of the day a senior brand manager is the point person who leads the efforts, pulls together the plan, gets it on paper, coordinates all team members, and delivers the presentation.

How Do You Write a Global Plan?

If you are writing a global brand plan then you need to obtain input and include various geographies. A best practice is to include on your core team a few members from three to five key countries that will account for about 80 percent of your business. Keep in mind that for the vast majority of brands, the US affiliate represents the majority of dollar sales and the vast majority of profit—however, not necessarily volume. Making sure your brand will be successful in the US market is paramount.

We then follow the 80/20 rule, which has two aspects. First, we try to make the plan work and be perfect for 80 percent of our business. It will not be perfect for all markets. However, if you have the key markets on your core team, if you have listened to and incorporated input from the key geographies, and if you have found the common ground (things that are similar versus different), then you will develop a plan that will be successful. The second aspect of the 80/20 rule is that the various countries need to accept, align, buy in, and use 80 percent of what the global team has developed in the plan. The other 20 percent can be localized to various market nuances, customs, practices, competitors, and regulations.

When Do You Write the Plan and How Often?

The brand plan should be written every year, for a two-year period. Too many firms start this process way too early. There are two issues with that. First, the sooner you start, the longer you work on it. You will fill the time. Many firms spend the entire year just planning, which leaves little to no time for implementation. Second, if you start too soon in the year, then you do not have any results from the previous brand plan. It is impossible to tell if your plan is working or not, if your objectives are being met, if your game plan was successful, if your solutions are working. Should you keep the same tactic, revise it, or kill it? If you start your brand plan process in March and your tactics only started in January, it is impossible to fully evaluate their effectiveness. We recommend that your brand plans be written in the three-month period (at most) leading up to the approval of the plans and budgets.

What Format Should the Plan Be Written In?

Many brand plans are written in a Microsoft Word document. It is often easier to tell a coherent story if the plan is first written in Word and only then transferred into PowerPoint to tell the story in presentations. Often the written version is then sent out as a pre-read for the PowerPoint presentation. We find that a best practice is to do both. Writing out the story is an excellent way to crystalize it, ensure it provides context (which is often lost on a slide), and allow for a pre-read, which is incredibly helpful for the non-marketing people in the audience. The PowerPoint is helpful for the "big show" presentation in front of leadership. In either case, you need to find the story in your brand.

If you find yourself in a global role and are overseeing affiliates (various countries) around the world, then we suggest a few adjustments. For many affiliate employees, English is not their native language. We have found that asking nonnative-English-speaking affiliates to write brand plans in Word is an unnecessary burden. Not only is it a huge time suck, but it creates unneeded stress for very minimal value. We typically ask our affiliates to prepare a PowerPoint presentation only.

Brand Plan Rx Cohesion Map

As you have progressed through this book, you have methodically gone through the brand plan process, developing your brand plan step by step in a logical manner. In each chapter you have refined your understanding, come to conclusions, and made decisions. It is now time to pull all of those decisions together and to ensure that they are aligned, make sense, and have a "golden thread" holding them together from beginning to end. Your brand plan needs to be cohesive—it needs to be consistent and interconnected. The Brand Plan Rx Cohesion Map is designed to help you do just that.

In fact, we believe that the Brand Plan Rx Cohesion Map is such a critical element that we recommend beginning and ending your

BRAND PLAN Rx COHESION MAP

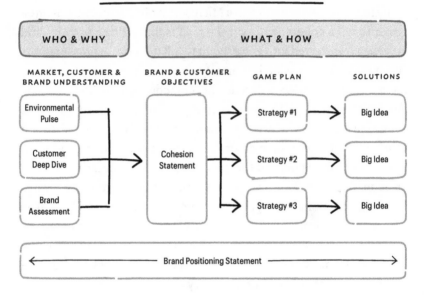

brand plan and presentation with it. It should be the first slide after introductory information and the last slide that you share with your audience. We like to begin with the Brand Plan Rx Cohesion Map because it lays out the entire plan. It tells the story in a simple manner, shows clarity in one snapshot, and highlights your "golden thread"— how things are aligned and fit together. It is a nice outline of what is to come and shows your thinking from soup to nuts; you literally are beginning with the end in mind. And we recommend the cohesion map as the last slide because it is a great summary that nicely recaps all of your analysis and decisions and is easy to remember. It is also a great set-up slide for your Q&A session, much better than one that only says "Q&A," "Thank You," or "Questions?" You have limited slides, and time, so don't waste one on a slide that does not provide significant value. Showing your Brand Plan Rx Cohesion Map as the last slide is much more productive, allowing for more discussion. This is the final impression you want your audience to take away.

You will notice that the components of the Brand Plan Rx Cohesion Map align to the steps in the brand plan process. It is no coincidence that this also aligns to your Brand Plan Rx Choice Map and to every chapter in this book. In each element and chapter you have made some key analysis, conclusion, or decision. You need to boil down the essence of each of these steps into one simple and clear statement or phrase that is then applied to the Brand Plan Rx Cohesion Map.

We have made many of these components incredibly simple, as you have already synthesized your thinking. For example, the positioning statement, the cohesion statement, brand objectives, customer objectives, and your strategy should all be very easy to take and transfer to the Brand Plan Rx Cohesion Map. You have already written them in the exact format that you can then transfer into the appropriate section in the Brand Plan Rx Cohesion Map. This should be simply a copy-and-paste exercise.

The environmental pulse, customer deep dive, brand assessment, and big idea might require a bit more thought and work to summarize appropriately. You should have distilled the analysis and thinking at this point; however, it might not be succinct or memorable enough to nicely fit on the Brand Plan Rx Cohesion Map. This might require more thinking and decision making. For example, the environmental pulse and customer deep dive might require you to summarize your work into three or four key bullet points.

Once you have finished transferring each section from each chapter and have completed the Brand Plan Rx Cohesion Map, the next and final step is crucial. You need to step back and take an objective look. Does this all make sense? Is it *cohesive*? Do a critical and objective review of the analyses, conclusions, and decisions you have made. This is the ultimate test of your brand plan—does it all make sense and is it compelling, based on an interconnected rationale of how your brand will soar?

Here are some ways to test your cohesion map and ensure that it is done well and achieves its purpose:

- **Target patient:** Do you see the same key target patient in the entire Brand Plan Rx Cohesion Map? More specifically, is the environmental trend you selected a key trend that is meaningful to your target patient? Are your solutions designed around your target patient?

- **Positioning:** Does your entire plan align to your positioning? Can you see the relevance of the key emotional benefit of your positioning in the other elements of the cohesion map?

- **Insight:** Does the insight in your customer deep dive show up in your strategy and solutions?

- **Solutions:** Will your solutions, if implemented correctly, meet your brand and customer objectives?

- **Working it backwards:** Start with a key solution you have developed (the last step) and literally go backwards through your brand plan. Does it align to your strategy? Does it achieve your brand and/or customer objectives? Does it fit into one of your four key decisions? Does it support your positioning? Are you leveraging your brand's superpower? Does it act on a key insight you discovered in your customer deep dive? Are you picking up on a tailwind or headwind that you analyzed in your environmental pulse?

There are endless ways to test your Brand Plan Rx Cohesion Map and play out various scenarios. Validating it in this manner is time well spent. Not only will your cross-functional team and leadership do this but, more importantly, your customers will in the marketplace. In the real world, patients and providers will be testing and validating your brand plan on a daily basis. If it all works and hangs together, then your brand will be successful, you will meet your business objectives, and you will have helped people live a better, healthier, and longer life. If your brand plan is not cohesive, the marketplace will tell you loud and clear.

Assumptions, Risks, and Contingencies

Every plan has inherent risks. Even the best designed and thought-out plans carry some risk. In fact, one could argue that if your plan does not have risks, then you aren't trying hard enough. You aren't pushing the envelope of the real opportunity for your brand. You are not realizing the brand's true potential. You are not helping as many people as possible.

Carrying risk is a good thing as long as you have planned for it and have contingencies in place. Start your risk analysis by listing the key assumptions you are making. It is important to be explicitly clear on these. Make sure that they are stated and that you have alignment among the team and with senior management. These assumptions not only impact your objectives, game plan, strategy, solutions, and forecast but also influence what you would consider to be risks to your plan. Assumptions could include market growth and trends, competitive moves or launches, changes in FDA regulations, changes in payer status, publications of studies, and approval of a new indication or line extension (NILEX). These assumptions are major events that could impact your brand plan positively or negatively.

Now you are ready to consider the risks to your plan. Begin by listing things that might go wrong that could result in you not making your plan and achieving your brand's objectives. Consider internal and external factors. Consider controllable and uncontrollable. Consider short term and long term. Assign probabilities on the likelihood that these events will happen. For each one, consider the impact on your business if the event were to happen; would the occurrence of that event have a huge impact on your business, or would it be manageable? If you chart these on a grid, you can easily tell which of these events are most likely to occur and which will have the greatest impact on the brand or market.

For the top events in the "high" quadrant of the grid (highest likelihood and highest impact), develop contingency plans. For the rest,

EVALUATING RISK

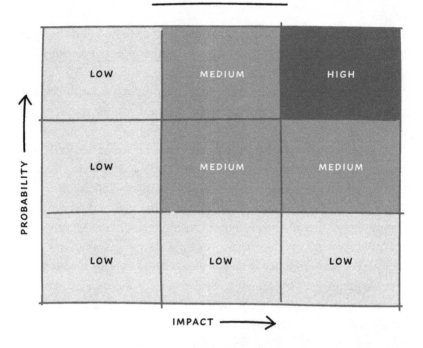

in the medium and low quadrants of the grid (low likelihood and low impact, high likelihood and low impact, and low likelihood and high impact), keep them in mind but don't spend any time planning for the eventuality. Your contingency plans specify what will be done should the event occur. Some can be steps that can be taken now, prior to the event occurring, to mitigate the risk. You should also plan for steps that are taken when the event occurs. You might also come up with measures that can be taken after the event occurs. Consider these as backup plans.

Often the contingency plans have dollar amounts attached, to pay for implementation. Think of this in two ways. First, will your action plan require funding? This is fairly straightforward; a certain action will cost money. The second concept is what you should do with the operating expenses you have associated with a solution that is directly tied to an event. Should the event occur, do you need to spend more or less of your planned budget? The natural inclination is to stop spending money on a solution should an unfortunate event occur. That might not always be the best course of action. For example, let's say your brand receives a black box warning. Some might think that you need to scale back on sales force promotion, because it will be harder to sell the brand and you are looking at a significant loss in sales. Another way to look at this is to consider what would happen if you "invested" in this risk—in other words, proactively managed it. In the example of the black box warning, you would increase activities with more promotion, social media spend, and direct-to-consumer (DTC) commercials, and/or by hiring more salespeople to help communicate and inform prescribing doctors about the situation, thereby minimizing the negative impact.

Timeline and Milestones

Developing a timeline that covers the length of your brand plan will help you determine the plan's feasibility. Your timeline should cover what needs to happen, and when, in order for the plan to work. It should identify key events and solutions with the relevant dates. Not only is it nice to see how everything fits together and to see it all on a timeline; it is also useful for a number of other reasons. It will help you either feel confident that everything is feasible or see that it is mission impossible. You might see that all of the major events requiring time, people, and resources are happening during the same month. You might find out that a solution you need to start at the beginning of your planning period will not have the data or publication available till the end of your planning period. As a result of this exercise you might find that you'll need to adjust the timing of some of your solutions.

In turn, you will also need to adjust the timing of sales and expenses associated with that solution.

Setting up and identifying key milestones helps to ensure that things are on track and that there are no surprises at the end of the year. The milestones also help you to recognize if and when things might get off track and allow you the opportunity to put things back on track. Some milestones might be obvious or more natural: the launch of a competitor, the publication of key data, or a NILEX. With others you might need to be a bit more curious and creative—examples of these include changes in FDA regulations, changes in access or reimbursement, and adjustments in public health policy like Medicare and Medicaid.

Finally, some milestones might need to be fabricated. This helps in creating opportunities as well as identifying triggers. An example is celebrating a first post-launch anniversary, or the number of patients served: one million patients or 100,000 lives saved. You could also take advantage of holidays, as Cialis did by creating events around Valentine's Day. Setting up milestones as triggers is a great way to hedge your bets, obtain additional funding, or secure/defend resources, if, for example, you need an additional $500,000 to run a DTC campaign associated with a new delivery device and you do not have the funds in your base budget. A way to negotiate for the additional expenses is to tie the release of those funds to approval and launch of the device. In other words, your plan is that if the event is imminent then you will ask that the funds be triggered and released.

Financials and Metrics

No brand plan is complete without a financial analysis. This is a tricky topic, as it could become a slippery slope and the brand plan could very quickly turn into a budget review. Senior executives are often highly interested in the dollars and cents. This is especially likely if the executives are not professional marketers, since they might feel uncomfortable in the qualitative, touchy-feely aspects of marketing.

To maintain control of the conversation while at the same time effectively communicating the financials, we suggest that you construct a profit and loss (P&L) that is fairly high level. By that we mean forecast out the monthly dollar sales over the brand plan time frame, show the associated share of market (SOM) and units, and then outline your monthly operating expenses to include programs, compensation and benefits (C&B), and cost of goods sold (COGS). The net is your high-level profit.

You will need to set up some metrics associated with your brand plans. Each solution should have an associated metric so that you can tell if you are doing well or not. You will need to answer these questions: Is this program delivering the way we thought? Are we getting our return on investment (ROI) from the program? Examples of these metrics include NPS (net promoter score), SOM, unaided awareness, TRx (total prescriptions), NBRx (new to brand prescriptions), and number of refills. A great pressure test for your solutions is to ask a series of questions:

1 What is the best solution that we believe most strongly will drive our results?

2 What would happen if we increased our efforts (spend, frequency, depth, etc.) on this solution tenfold?

3 How much more sales or SOM could we obtain if we doubled down on our best solutions?

It would be a good idea to run these scenarios prior to your brand plan review. Management might ask you to do this, but they will also likely ask you to fund the solutions from your other solutions that you did not identify as your best. It is unlikely that they have a pot of gold and are handing out extra operating expenses.

Storytelling

Now that you have completed your analysis, thinking, planning, and decision making, it is time to put everything together into one compelling and cohesive story. Writing a brand plan is all about telling a story. Stories are memorable, impactful, and inspiring—everything you want your brand plan to be. A better story is more likely to be approved, gain alignment, and, most importantly, be implemented well.

Storyboarding: The best way to build your brand plan is to start with a storyboard. Storyboarding is a simple three-step process:

1   On a whiteboard or PowerPoint, create "empty" slides with only the subject. What will this slide be about (e.g., title page, agenda page, cohesion map page, positioning page)? At this stage don't worry about the content, data, or information.

2   For each slide, then write out your header and kicker—the main point and the main takeaway of the slide, respectively. The header is the statement that goes across the top of the slide, above the line (body of the slide). It states the *claim/statement* that the slide is making. For example, "Telehealth usage among patients with diabetes has grown 250% in the past year." The body of the slide should show the charts and graphs to support that claim. The kicker is the *conclusion/takeaway* that you want your audience to have a result of the information and data. For example, "We need to promote our brand to telehealth companies and prescribers."

3   In time you will begin to fill in the content of the slide. This is the information—the charts, the graphs, the quotes, the analysis, etc.— that supports your header and kicker. In our example, this could include a chart that shows the growth in telehealth and a chart that shows your brand's spend in telehealth promotion versus that of

your competitors. Throughout the process you need to step back and look at your storyboard to ensure that the pages flow seamlessly and logically, to identify what data gaps exist, and to see if you like the story that you are telling. If you took all of your headers, listed them on a page, and read them from beginning to end, would they make sense, and do they tell your story?

Theme: It is a good idea to create a theme that envelops the heart of your plan. It should capture the main idea, be memorable, and apply to the situation. It should also be relevant and something that everyone can rally behind. The following are some examples: Building a new category, Transforming care, Bringing innovation to market, Simple solutions, Breathing easier together, Saving millions of lives.

Like all good stories, yours should have a beginning, middle, and end. A good way to frame this is to remind your audience what the current state of the brand is. Then tell them what your plan is. End with what the future holds for the brand if the plan is accepted, approved, funded, and implemented. Find a way to connect different pieces of information and create a coherent, logical, simple, and aligned narrative flow that ties the entire situation together.

Brand plan stories don't use the same narrative techniques as novels. A best practice in the business world is to say what you are going to say, say it, and then say what you just said. Your story should do this. Also, it is best to begin with the end in mind. Executives do not need a big build-up with drama and suspense to only find out the key points are at the very end. No need to keep anyone on the edge of their seat. This only builds frustration, raises too many questions, allows for things to go catawampus and possibly derail, and wastes precious time. Get to the point. Especially if you have some bad news, such as that the past year's performance is below expectations, a new competitor has made more progress than you expected, or new regulations will make things more challenging than forecasted. And if you do have bad news, use the sandwich approach: tell them something good, tell them the bad news, and then finish with something good. Don't think of it as sugarcoating reality or hiding the truth; it is more about keeping things in context.

A good story will summarize the entire plan in three or four easy-to-follow sentences (as bullet points). These should come from and be exactly what you have on your Brand Plan Rx Cohesion Map.

For example:

- The cardiovascular market is unsatisfied, patients are frustrated with current treatments, and doctors wish they could save more patients with a product that had better efficacy. As a result, according to the National Center for Health Statistics, ten million patients are undertreated and two million die unnecessarily each year in the United States.

- Our brand has the ability to truly transform the marketplace with a novel mechanism of action (MOA) that provides better efficacy, allowing doctors to treat and save 10 percent more patients and extend life by two years.

- We believe we can obtain a 15 percent SOM, which will generate $1 billion in sales by the third year of launch.

- To achieve this, we need $300 million to create and grow a new category by introducing the novel MOA, build the market by creating awareness, develop a DTC campaign, and hire a sales force to call on the top one thousand prescribers. Together, we can truly transform the treatment of this disease and help save lives.

Think of this as your elevator pitch that can be repeated by anyone, anywhere, at any time.

Your story needs to answer the five Ws plus one H: *who, what, when, where, why,* and *how.* And not in the traditional order.

- **Who:** segmentation and targeting

- **Why:** insights

- **What:** objectives

- **How:** game plan

- **When and Where:** solutions, the big idea

Slide deck: Your presentation and slide deck should be as short as possible. You want to tell your story briefly and leave as much time as you can for discussion. It is during the Q&A, after your formal presentation, that the real understanding, buy-in, and agreement happen.

Your plan should not be more than twenty slides. Too often brand teams come to the brand review with over a hundred slides to tell their story. We understand—you have probably created at least that many slides—but it is not productive. It shows that you haven't made the hard choices and haven't pulled together a cohesive strategy. It is hard to leave many of those slides out of the presentation, but it is much more important to show command of your brand by displaying that you are able to crystalize your thinking rather than bombarding your audience with a hundred slides and asking them to synthesize what you have presented. You are much better off controlling the message than letting your audience draw their own conclusions. The Brand Plan Rx Cohesion Map, which also serves as your executive summary, will help you to keep the number of slides to a minimum. Keep the other eighty slides as backups and use them for the Q&A and discussion part of the presentation.

Here is a suggested format:

SLIDE 1	Title page, brand plan name, date, presenters, location, and theme
SLIDE 2	Brand plan agenda
SLIDE 3	Brand Plan Rx Cohesion Map
SLIDE 4	State of the brand and executive summary: performance, results versus plan, challenges and opportunities
SLIDE 5	Environmental pulse
SLIDES 6–7	Customer deep dive: key insights and target patient, patient journey and MOTs
SLIDE 8	Brand assessment: FAB and superpower
SLIDE 9	Positioning statement
SLIDE 10	Brand objectives: cohesion statement
SLIDE 11	Game plan: three strategies and key solutions
SLIDES 12–16	Solutions: one to two pages for the top two or three tactics that will drive your business
SLIDE 17	Assumptions, risks, and contingencies
SLIDE 18	Timeline
SLIDE 19	Financials and metrics
SLIDE 20	Conclusions: Brand Plan Rx Cohesion Map (Slide 3 repeated)

Sources, References, and Market Research

It is important to source and reference all of your data and claims. References add credibility. Your own market research is a reference and still a valuable source. The source should be footnoted on each slide.

A best practice is to start off your presentation with three to five key and relevant facts from credible sources in your executive summary. This will set the tone for the rest of the presentation, gain attention, and strongly reinforce the points you are making.

Key Questions

- Do your elevator pitch and your cohesion map align?

- What challenge questions you can anticipate? How would you answer them? Who is the best person on the team to answer various questions?

- If you were given more budget, how would you spend it?

- If some of your budget were taken away, what would you cut?

- How do you think the competition would react to your plan? What response in the marketplace do you anticipate?

- How does this compare to last year's plan? What has changed?

- Are your timelines too aggressive? Or not aggressive enough?

CONCLUSION

Key Learning Points

Key Takeaways: Choice and Cohesion

Nemawashi

Power of Questions

Lessons Learned

Modern Marketing: Parting Thoughts

BRAND PLAN Rx CHOICE MAP

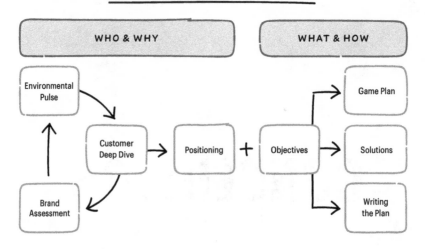

BRAND PLAN Rx COHESION MAP

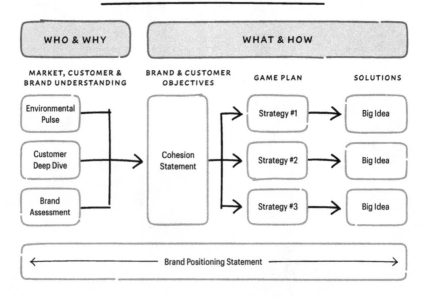

Key Takeaways: Choice and Cohesion

The most important takeaways from this book are the two tenets that we first mentioned in the introduction.

1 **Choice:** If you can't summarize a brand plan on one page, you haven't made the tough decisions to drive success. The only way to obtain that simplicity is by making clear choices.

2 **Cohesion:** It all has to fit together. There should be a "golden thread" that aligns the plan to a common understanding. Your analysis should lead to clear choices and your corresponding strategy and tactics should all be consistent.

In order to achieve both, we have two constructs that help you as you are developing a brand plan: the Brand Plan Rx Choice Map and the Brand Plan Rx Cohesion Map.

A brand plan is a powerful tool. It is one of the most important documents your firm will produce, where you select key aspects from all your discovery of your choice map to frame your story. For a brand manager it is the be-all and end-all. This annual exercise is not only a health check on your brand; it also sets the course for the next two years. Without an aligned, realistic, inspiring, growth-oriented plan, your brand is sure to not meet its objectives. The brand plan gives a clear update on the current situation, anticipates trends, and articulates who your customer is, what insights you have garnered, what position you want to obtain in your customer's mind and heart, what goals and objectives you hope to achieve, and the solutions and strategic initiatives that will lead you to them. It specifies the tactics, expenses, timelines, risks, and metrics that need to be implemented. A brand plan states where you are going, why you are going there, and how you will get there.

Nemawashi

We would like to borrow an idea and concept that has been mastered in Japan: *nemawashi*. This is an informal process of quietly laying the foundation for your brand plan by talking to various stakeholders and gathering their input, support, and feedback. It is literally bringing people along in the process so that when it comes time for approval, no one is surprised or sees things for the first time. It will also ensure that you are not ambushed by someone's public dissent. It goes a long way in ensuring true buy-in. In other words, the more you involve me, seek my input, and listen to and incorporate my ideas, the more I will be behind your plan.

To build a consensus by *nemawashi*, you need to make a list of key stakeholders and team members. Then set up time with them to share your brand plan and listen to their ideas—and then do your best to incorporate them. Include as many stakeholders as is practical in your workshops as well.

Power of Questions

Asking the right questions is a key element in ensuring that you have a well-thought-out and effective brand plan. It is also a great way to drive cohesion. The answers to these key questions are essential, and this book shows you how to reach them. If you are able to answer each chapter's questions for your brand, then without a doubt you will have developed a plan that will allow your brand to reach its full potential and even exceed expectations.

EE Cummings suggested that there always are beautiful answers, but the one who asks a more beautiful question enables a better answer. In our journey to consulting and driving brand plans, we have found that effective brand plans begin and end with insightful questions. You will note that at the end of each chapter we have included "money" questions. We have focused on practical questions that enable you to critically assess your decisions and how each critical part of your brand plan plays a symbiotic role. Depending on your therapeutic category, these questions may change. Taking a step back and interrogating the plan you have put together is paramount.

What do you think are the most difficult questions someone could ask you about your brand? If you were the CEO of the organization, what would you ask before deciding when and why you would invest? It is a good idea to schedule a role-play scenario with someone you trust who is not close to the plan. Ask them to really challenge you, so that you are uber-prepared for any question that may come up.

Some of our favorite questions are the following:

- How confident are you that you can achieve the objectives you have laid out, and why?

- If I doubled your forecast, what budget would you need and what big ideas would you employ to achieve that target?

- How will you ensure that you will surpass the competition?

- Why should we invest in this franchise over another? It seems that we aren't the leader, so is it worth us going all in?

- What did you learn from last year that you applied to your decisions for this year?

- What did you learn from your competitors, or the marketplace, that informed your plan?

Lessons Learned

Together we have seen the brand plan process play out thousands of times. *Brand Plan Rx* is full of lessons that we have learned based on success and failures. Often, the tendency is to focus on the failures, what could go wrong, avoiding mistakes, and playing it safe. However, we prefer to play to win. We have seen incredible brand plans that have galvanized an organization, transformed a marketplace, exceeded expectations, and, most importantly, helped to make millions of people's lives healthier and better. Across these wildly successful brand plans we have noticed a couple of things that they all consistently do.

- **Align your team:** We highly recommend that you get your entire team aligned with the brand plan—and not just your marketing team; include cross-functional colleagues in medical, regulatory, sales, manufacturing, market research, and other areas. In addition, include your partners, vendors, and agencies. The best way to get everyone aligned is to involve them early, listen to and value their contributions, and have them read *Brand Plan Rx* so that you are working with the same knowledge.

- **Be bold:** Think big, think innovatively, be curious, and go for it. When you think about what solutions are going to fully realize your game plan, don't hold back. Look for an idea that would move the patient or doctor, be valuable in their lives, and earn the role your brand is striving for.

- **Less is more:** It's always better to have a few areas of focus versus a litany of ideas. If you can focus your energy and investment where it counts, you will achieve more.

- **Place your bets:** Be clear about where you are placing your bets and why, the assumptions, and the upside and downside of those assumptions. This will require you to know your market, the environment, and your brand inside and out. Understand the headwinds and tailwinds, squeeze on economic pressure, new technology, data, consumer choice models. There will always be change in a marketplace, whether minimal or substantial. But it's your job to make a tough call based on the data and your understanding of the marketplace. You can't predict everything, but you can explain your rationale and be clear about where and why you are focusing your game plan in a certain direction. Informed "bets" can make or break your brand.

Modern Marketing: Parting Thoughts

We have talked about how COVID-19 accelerated the need for a modern marketing approach. In fact, COVID-19 did us a favor (we realize that's a bit of a controversial suggestion). The pandemic is a portal through which to look at how we can serve consumers differently. For umpteen years the health and wellness sector has operated in a siloed, conventional model, not thinking of the consumer or seeing the market as a living, breathing ecosystem. As we have pointed out, healthcare is ripe for disruption. Some of the modern marketing thinking that has been accelerated as a result of COVID-19 has forced us to consider the following:

1 **Ecosystem thinking:** Look at how we need to see the system of care and its consumers holistically. Let's not just solve the patient's problem, the doctor's problem, or the managed care value proposition; let's look at how they all interrelate and can simplify a solution for all parties. We need to truly work together, by aligning incentives within the healthcare ecosystem. We have seen this happen with COVID-19 vaccines and treatments. All sectors of the industry are collaborating at an unprecedented level, creating a new paradigm in partnerships.

2   **Technology:** Embrace and anticipate technology as a friend, not a foe. How can interoperability of electronic health records (EHRs), care path AI intelligence, or omnichannel marketing drive better outcomes for all stakeholders in the ecosystem?

3   **Place of care:** The place of care is evolving. When and where doctors are connecting with patients and providing care is changing dramatically. The concept of "care anywhere" has become a reality as a result of COVID-19. When and where doctors are learning about new diseases and treatments has also changed. Where you get a second opinion and how is different. It is all dynamic. So think about flexibility and bespoke ways to drive value to all stakeholders. If waiting rooms are a thing of the past, think about the car as a place where a patient can learn more about their disease and formulate clear questions before they speak to their doctor.

4   **Science matters:** Science matters more than ever, and it's moving at the speed of light. Doctors, health systems, and centers of excellence all need to be apprised of new, cutting-edge ways of treating, diagnosing, translating learning, empathizing, and delivering care. COVID-19 has reinstituted the health expert as the expert. It is nearly impossible to find the time to keep up with all the advances in science and technology. Think about new ways for these stakeholders to stay abreast of what matters. For example, curation of real-time discovery will be a competitive advantage, not just about your product but about the category and the industry. And if doctors are attending virtual symposia, not only from experts but from patient key opinion leaders (KOLs), think about data visualization and translation of the data. How can we ensure health and science literacy among health experts and simplify clearly so they can translate to their patients?

5   **Applying data:** Data is at the center of your strategy. Not clinical data only, but human data—how people make decisions. What is

your company's data strategy and philosophy? How do you harness and digest data in real time to learn as much as you can about how people feel about their disease, treat patients, and manage the disease so that we can solve problems holistically and with relevance? Buying or otherwise acquiring data and synthesizing and integrating it in a compliant, transparent way will enable you to have the pulse of your customers and the initiatives you employ.

6 **Real-time learning:** Institute real-time data learning and feedback versus waiting for quantitative ATU (awareness, trial, usage) studies to find out how your brand is making connections. As in other industries, make changes in real time.

7 **Start-up mentality:** Start-ups are agile. If you are working for a larger company with many components, it's difficult to institute a way of working that sets up success for innovation and invention. Think about how you can create that culture, subteam, or working group to employ tests and learn models. Think through many ways of connecting, and experiment with the information, message, and initiatives on archetypes of consumers, doctors, and health extenders, to see what's most valuable.

8 **Purpose-driven marketing:** People care about values now more than ever before. COVID-19 has created skepticism and uncertainty. Many people want to know that the company they are buying their product from has a value set that matches their own. This is provocative, as medicine is something we need rather than want or choose. We are starting to see that this mentality applies to health products as much as to any consumer good. If the company is invested in me, in me getting better, in me serving the patient better, it matters. There may be big ideas beyond just the medicine to demonstrate that your company cares about the ultimate outcome versus just selling medicine and services.

9 **Authenticity and underserved populations:** The COVID-19 pandemic has exposed the importance of healthcare to all, as well as the lack of consistency and effort in prevention, understanding, and treatment for underserved groups. Less access to care, lower quality of care, and lower health literacy all drive poorer health outcomes. It is our responsibility as marketers to think through how to earn trust and build brands with positive outcomes in mind across all audiences who could benefit from our brand. One universal truth to your efforts in this regard is authenticity. The only way to authentically connect is to ensure your teams and your agency partners represent the population you are engaging with.

10 **Corporate equity and brand equity:** A related point to purpose-driven marketing is understanding the importance of corporate equity and brand equity. More and more consumers and doctors care who is behind the medicine and services, what their values are, and how they uphold those values and earn trust. Think about equity in a new way: Is it brand equity you need to drive, or is the company behind the brand important to the consumer, health system, and provider? We are seeing that both matter. The actions and decisions of the brand and the company together can drive trust, loyalty, and recommendations.

11 **Definition of a brand:** What defines a brand? Who is the brand? That is, is it the sum of its clinical benefits, or the equation of the experience connected to the disease and how the patient receives the brand and pays for the medicine? Throughout the entire touchpoint of the care path this matters; from awareness of disease to contemplation to fulfilling the prescription and refilling the prescription, the experience matters. The brand equals the experience tethered to it.

12 **Building or partnering for innovation:** When thinking about how to innovate, approach problems differently. Many companies have felt they needed to build expertise internally to drive innovation. Sometimes this has advanced their impact, but sometimes the investment has been lost. Partnering is an important consideration. If AI, machine learning, and data automation are new entities for your team and organization, partnering with a company that is an expert can accelerate your go-to-market innovation. It will be critically important to upskill your internal team or bolster your team with critical expertise as you partner, so you can assess the value of the company and ask the tough questions to get the most out of the partnership relationship.

We have created the Brand Plan Rx Choice Map and Brand Plan Rx Cohesion Map so you have a clear path to structure your thinking. Sometimes the toughest job is to agree on the format, or structure, of the brand plan. With the components laid out in *Brand Plan Rx*, you can focus on the content and not get bogged down in what inputs are necessary to drive critical thinking for your brand.

We trust that *Brand Plan Rx* has given you the direction, ideas, tools, and constructs you need to build a cohesive brand plan that makes clear and bold choices, all with the aim of helping people live longer and healthier lives while at the same time achieving your brand's business objectives.

NOTES

1 George S. Day and Christine Moorman, *Strategy from the Outside In: Profiting from Customer Value* (New York: McGraw-Hill, 2010), 5.

2 Physician age ranges: Statista, "Distribution of Physicians by Age Group in 2018," statista.com/statistics/415961/share-of-age-among-us-physicians/; physician burnout study: Medscape, *Medscape National Physician Burnout & Suicide Report 2020: The Generational Divide* (New York: Medscape, 2020), medscape.com/slideshow/2020-lifestyle-burnout-6012460#1.

3 "Drug Overdose," Drug Policy Alliance, drugpolicy.org/issues/drug-overdose.

4 "Heart Disease Facts," Centers for Disease Control and Prevention, last reviewed September 8, 2020, cdc.gov/heartdisease/facts.htm.

5 Riley Griffin, "Two Big Drug Flops Show How Health-Care Economics Have Changed," *Bloomberg Businessweek*, January 9, 2020, bloomberg.com/news/articles/2020-01-09/these-big-drug-flops-show-how-healthcare-economics-have-changed.

6 As cited in Josh Bloom, *Whatever Happened to AIDS? How the Pharmaceutical Industry Tamed HIV* (New York: Special Report by the American Council on Science and Health, 2011), 14.

7 Manisha Patel et al., "National Update on Measles Cases and Outbreaks—United States, January 1–October 1, 2019," *Morbidity and Mortality Weekly Report* 68, no. 40 (October 11, 2019): 893–96, cdc.gov/mmwr/volumes/68/wr/mm6840e2.htm.

8 "Pertussis Cases by Year (1922–2018)," Centers for Disease Control and Prevention, last reviewed August 7, 2017, cdc.gov/pertussis/surv-reporting/cases-by-year.html.

9 Oleg Bestsennyy et al., "Telehealth: A Quarter-Trillion-Dollar Post-COVID-19 Reality?" McKinsey & Company, May 29, 2020, mckinsey.com/industries/healthcare-systems-and-services/our-insights/telehealth-a-quarter-trillion-dollar-post-covid-19-reality.

10 Elie Ofek and Ron Laufer, "Eli Lilly: Developing Cymbalta" (Harvard Business School Case 507-044, November 2006), store.hbr.org/product/eli-lilly-developing-cymbalta/507044?sku=507044-PDF-ENG.

11 For more on the Cialis story, see "Building More Loving Relationships in Commerce," ?What If!, whatifinnovation.com/case_studies/cialis/.

12 See Amy M. Barrett et al., "Epidemiology, Public Health Burden, and Treatment of Diabetic Peripheral Neuropathic Pain: A Review," *Pain Medicine* 8, no. S2 (September 2007): S50–62, doi.org/10.1111/j.1526-4637.2006.00179.x.

13 "The Z-Pak idea came from a Pfizer marketing executive who had previously worked in the birth control market and had firsthand experience on how a packaging solution had improved actual and perceived value of a drug. The Z-Pak package instructed patients to take pills one and two on the first day, and another pill for each of the next four days. Patients complied nearly 100% in taking the full dose. Compare this to Biaxin, which was a total of two doses over ten days—or twenty chances to either miss a dose, skip a dose, or stop taking it once you started feeling better. Interestingly, numerous studies proved that the efficacy of Biaxin was superior to that of Zithromax, and market research indicated that doctors knew and recognized that Biaxin was a more effective antibiotic. However, Z-Pak gave Zithromax a superpower that superseded all other product attributes. Within three years of its launch, Zithromax had reversed the old market ranking and was out selling Biaxin by four to one." Julie Hennessy, "Zithromax Z-Pak and the 'Biaxin BBQ' (A)" (Kellogg School of Management Cases, Northwestern University, 2017), doi.org/10.1108/case.kellogg.2016.000423.

14 Timothy Calkins and Joshua Neiman, "Crestor" (Kellogg School of Management Case 5-306-506, Northwestern University, 2006), 5, Table 1, kellogg.northwestern.edu/faculty/research/researchdetail?guid=6238e9f8-df90-49f7-8bdf-e9193ffe8329.

15 For more on Nexium, see Lynda Sears, "Unleashing the Power of
    Nexium," *PharmaVoice*, July 2001, pharmavoice.com/article/2001-07-
    unleashing-the-power-of-nexium/.

16 Tim Calkins, *Breakthrough Marketing Plans: How to Stop Wasting Time
    and Start Driving Growth*, 2nd ed. (New York: Palgrave Macmillan, 2012).

17 "SGLT2 inhibitors are a class of medications used to treat type 2 diabetes.
    They're also called sodium-glucose transport protein 2 inhibitors or
    gliflozins. SGLT2 inhibitors prevent the reabsorption of glucose from
    blood that's filtered through your kidneys, therefore facilitating glucose
    excretion in the urine." Heather Grey, "Everything You Wanted to Know
    about SGLTs Inhibitors," *Healthline*, last updated September 3, 2019,
    healthline.com/health/type-2-diabetes/sglt2-inhibitors.

18 Clayton M. Christensen, Jerome H. Grossman, and Jason Hwang, *The
    Innovator's Prescription: A Disruptive Solution for Health Care* (New York:
    McGraw-Hill, 2008).

19 See "News on Millennium Development Goals," We Can End Poverty:
    Millennium Development Goals and Beyond 2015, United Nations, n.d.,
    un.org/millenniumgoals/; "The 17 Goals," Sustainable Development,
    Department of Economic and Social Affairs, United Nations, n.d., sdgs.
    un.org/goals.

20 Eric Topol, *The Patient Will See You Now: The Future of Medicine Is in Your
    Hands* (New York: Basic Books, 2015), 166.

21 W. Chan Kim and Renée Mauborgne, *Blue Ocean Strategy: How to Create
    Uncontested Market Space and Make the Competition Irrelevant* (Boston:
    Harvard Business School Press, 2004).

ABOUT
THE AUTHORS

MARKUS SABA is a marketing professor at the University of North Carolina Kenan-Flagler Business School and the executive-in-residence of the UNC Center for the Business of Health. An experienced pharmaceutical executive with expertise in global and health-care marketing, Saba held numerous market-ing leadership positions during his twenty-
five years with Eli Lilly and Company, building some of the most suc-cessful brands in the industry. He received his MBA from the UNC Kenan-Flagler Business School and his bachelor of science in mar-keting from the Penn State Smeal College of Business. Through his consulting firm, Saba works with leading pharmaceutical, biotech, and healthcare providers, payers, and service companies.

HILARY GENTILE is the global chief strategy officer of McCann Health, where she has worked for more than twenty years. She drives strategy and innovation for McCann's teams across all of the company's agencies around the globe. Gentile advises McCann's clients across the spectra of the pharmaceutical and health and wellness industries,  employing her extensive experience in new product launches and business planning. As a graduate of Smith College in sociology and biology, her insights into the totality of the health ecosystem enable her to drive action plans for a variety of stakeholders. Gentile is a frequent keynote speaker for clients and industry events, on subjects ranging from the future of the health and wellness industry, to empathy-led healthcare marketing, to data-driven strategies for wellness. In 2017, Gentile was honored to be inducted into the MM&M Hall of Femme. She lives in Berkeley Heights, New Jersey, with her husband, Tom; son, Gregory; daughter, Vittoria; and dog, Jake.

CPSIA information can be obtained
at www.ICGtesting.com
Printed in the USA
BVHW080455190321
602956BV00001B/88